T0073962

LEADERSHIP, MANAGEMENT AND CULTURE FOR SAFETY IN RADIOACTIVE WASTE MANAGEMENT

The following States are Members of the International Atomic Energy Agency:

AFGHANISTAN	GEORGIA	OMAN
ALBANIA	GERMANY	PAKISTAN
ALGERIA	GHANA	PALAU
ANGOLA	GREECE	PANAMA
ANTIGUA AND BARBUDA	GRENADA	PAPUA NEW GUINEA
ARGENTINA	GUATEMALA	PARAGUAY
ARMENIA	GUYANA	PERU
AUSTRALIA	HAITI	PHILIPPINES
AUSTRIA	HOLY SEE	POLAND
AZERBAIJAN	HONDURAS	PORTUGAL
BAHAMAS	HUNGARY	QATAR
BAHRAIN	ICELAND	REPUBLIC OF MOLDOVA
BANGLADESH	INDIA	ROMANIA
BARBADOS	INDONESIA	RUSSIAN FEDERATION
BELARUS	IRAN, ISLAMIC REPUBLIC OF	RWANDA
BELGIUM	IRAQ	SAINT LUCIA
BELIZE	IRELAND	SAINT VINCENT AND
BENIN	ISRAEL	THE GRENADINES
BOLIVIA, PLURINATIONAL	ITALY	SAMOA
STATE OF	JAMAICA	SAN MARINO
BOSNIA AND HERZEGOVINA	JAPAN	SAUDI ARABIA
BOTSWANA	JORDAN	SENEGAL
BRAZIL	KAZAKHSTAN	SERBIA
BRUNEI DARUSSALAM	KENYA	SEYCHELLES
BULGARIA	KOREA, REPUBLIC OF	SIERRA LEONE
BURKINA FASO	KUWAIT	SINGAPORE
BURUNDI	KYRGYZSTAN	SLOVAKIA
CAMBODIA	LAO PEOPLE'S DEMOCRATIC	SLOVENIA
CAMEROON	REPUBLIC	SOUTH AFRICA
CANADA	LATVIA	SPAIN
CENTRAL AFRICAN	LEBANON	SRI LANKA
REPUBLIC	LESOTHO	SUDAN
CHAD	LIBERIA	SWEDEN
CHILE	LIBYA	SWITZERLAND
CHINA	LIECHTENSTEIN	SYRIAN ARAB REPUBLIC
COLOMBIA	LITHUANIA	TAJIKISTAN
COMOROS	LUXEMBOURG	THAILAND
CONGO	MADAGASCAR	TOGO
COSTA RICA	MALAWI	TRINIDAD AND TOBAGO
CÔTE D'IVOIRE	MALAYSIA	TUNISIA
CROATIA	MALI	TURKEY
CUBA	MALTA	TURKMENISTAN
CYPRUS	MARSHALL ISLANDS	UGANDA
CZECH REPUBLIC	MAURITANIA	UKRAINE
DEMOCRATIC REPUBLIC	MAURITIUS	UNITED ARAB EMIRATES
OF THE CONGO	MEXICO	UNITED KINGDOM OF
DENMARK	MONACO	GREAT BRITAIN AND
DJIBOUTI	MONGOLIA	NORTHERN IRELAND
DOMINICA	MONTENEGRO	UNITED REPUBLIC
DOMINICAN REPUBLIC	MOROCCO	OF TANZANIA
ECUADOR	MOZAMBIQUE	UNITED STATES OF AMERICA
EGYPT	MYANMAR	URUGUAY
EL SALVADOR	NAMIBIA	UZBEKISTAN
ERITREA	NEPAL	VANUATU
ESTONIA	NETHERLANDS	VENEZUELA, BOLIVARIAN
ESWATINI	NEW ZEALAND	REPUBLIC OF
ETHIOPIA	NICARAGUA	VIET NAM
FIJI	NIGER	YEMEN
FINLAND	NIGERIA	ZAMBIA
FRANCE	NORTH MACEDONIA	ZIMBABWE
GABON	NORWAY	

The Agency's Statute was approved on 23 October 1956 by the Conference on the Statute of the IAEA held at United Nations Headquarters, New York; it entered into force on 29 July 1957. The Headquarters of the Agency are situated in Vienna. Its principal objective is "to accelerate and enlarge the contribution of atomic energy to peace, health and prosperity throughout the world".

IAEA SAFETY STANDARDS SERIES No. GSG-16

LEADERSHIP, MANAGEMENT AND CULTURE FOR SAFETY IN RADIOACTIVE WASTE MANAGEMENT

GENERAL SAFETY GUIDE

INTERNATIONAL ATOMIC ENERGY AGENCY
VIENNA, 2022

COPYRIGHT NOTICE

© IAEA, 2022

Printed by the IAEA in Austria
January 2022
STI/PUB/1979

IAEA Library Cataloguing in Publication Data

Names: International Atomic Energy Agency.
Title: Leadership, management and culture for safety in radioactive waste management / International Atomic Energy Agency.
Description: Vienna : International Atomic Energy Agency, 2022. | Series: IAEA safety standards series, ISSN 1020–525X ; no. GSG-16 | Includes bibliographical references.
Identifiers: IAEAL 21-01463 | ISBN 978–92–0–137421–9 (paperback : alk. paper) | ISBN 978–92–0–137521–6 (pdf) | ISBN 978–92–0–137621–3 (epub)
Subjects: LCSH: Radioactive wastes — Safety measures. | Radioactive wastes — Leadership. | Radioactive wastes — Management.
Classification: UDC 621.039.7 | STI/PUB/1979

FOREWORD

by Rafael Mariano Grossi
Director General

The IAEA's Statute authorizes it to "establish...standards of safety for protection of health and minimization of danger to life and property". These are standards that the IAEA must apply to its own operations, and that States can apply through their national regulations.

The IAEA started its safety standards programme in 1958 and there have been many developments since. As Director General, I am committed to ensuring that the IAEA maintains and improves upon this integrated, comprehensive and consistent set of up to date, user friendly and fit for purpose safety standards of high quality. Their proper application in the use of nuclear science and technology should offer a high level of protection for people and the environment across the world and provide the confidence necessary to allow for the ongoing use of nuclear technology for the benefit of all.

Safety is a national responsibility underpinned by a number of international conventions. The IAEA safety standards form a basis for these legal instruments and serve as a global reference to help parties meet their obligations. While safety standards are not legally binding on Member States, they are widely applied. They have become an indispensable reference point and a common denominator for the vast majority of Member States that have adopted these standards for use in national regulations to enhance safety in nuclear power generation, research reactors and fuel cycle facilities as well as in nuclear applications in medicine, industry, agriculture and research.

The IAEA safety standards are based on the practical experience of its Member States and produced through international consensus. The involvement of the members of the Safety Standards Committees, the Nuclear Security Guidance Committee and the Commission on Safety Standards is particularly important, and I am grateful to all those who contribute their knowledge and expertise to this endeavour.

The IAEA also uses these safety standards when it assists Member States through its review missions and advisory services. This helps Member States in the application of the standards and enables valuable experience and insight to be shared. Feedback from these missions and services, and lessons identified from events and experience in the use and application of the safety standards, are taken into account during their periodic revision.

I believe the IAEA safety standards and their application make an invaluable contribution to ensuring a high level of safety in the use of nuclear technology. I encourage all Member States to promote and apply these standards, and to work with the IAEA to uphold their quality now and in the future.

THE IAEA SAFETY STANDARDS

BACKGROUND

Radioactivity is a natural phenomenon and natural sources of radiation are features of the environment. Radiation and radioactive substances have many beneficial applications, ranging from power generation to uses in medicine, industry and agriculture. The radiation risks to workers and the public and to the environment that may arise from these applications have to be assessed and, if necessary, controlled.

Activities such as the medical uses of radiation, the operation of nuclear installations, the production, transport and use of radioactive material, and the management of radioactive waste must therefore be subject to standards of safety.

Regulating safety is a national responsibility. However, radiation risks may transcend national borders, and international cooperation serves to promote and enhance safety globally by exchanging experience and by improving capabilities to control hazards, to prevent accidents, to respond to emergencies and to mitigate any harmful consequences.

States have an obligation of diligence and duty of care, and are expected to fulfil their national and international undertakings and obligations.

International safety standards provide support for States in meeting their obligations under general principles of international law, such as those relating to environmental protection. International safety standards also promote and assure confidence in safety and facilitate international commerce and trade.

A global nuclear safety regime is in place and is being continuously improved. IAEA safety standards, which support the implementation of binding international instruments and national safety infrastructures, are a cornerstone of this global regime. The IAEA safety standards constitute a useful tool for contracting parties to assess their performance under these international conventions.

THE IAEA SAFETY STANDARDS

The status of the IAEA safety standards derives from the IAEA's Statute, which authorizes the IAEA to establish or adopt, in consultation and, where appropriate, in collaboration with the competent organs of the United Nations and with the specialized agencies concerned, standards of safety for protection of health and minimization of danger to life and property, and to provide for their application.

With a view to ensuring the protection of people and the environment from harmful effects of ionizing radiation, the IAEA safety standards establish fundamental safety principles, requirements and measures to control the radiation exposure of people and the release of radioactive material to the environment, to restrict the likelihood of events that might lead to a loss of control over a nuclear reactor core, nuclear chain reaction, radioactive source or any other source of radiation, and to mitigate the consequences of such events if they were to occur. The standards apply to facilities and activities that give rise to radiation risks, including nuclear installations, the use of radiation and radioactive sources, the transport of radioactive material and the management of radioactive waste.

Safety measures and security measures[1] have in common the aim of protecting human life and health and the environment. Safety measures and security measures must be designed and implemented in an integrated manner so that security measures do not compromise safety and safety measures do not compromise security.

The IAEA safety standards reflect an international consensus on what constitutes a high level of safety for protecting people and the environment from harmful effects of ionizing radiation. They are issued in the IAEA Safety Standards Series, which has three categories (see Fig. 1).

Safety Fundamentals

Safety Fundamentals present the fundamental safety objective and principles of protection and safety, and provide the basis for the safety requirements.

Safety Requirements

An integrated and consistent set of Safety Requirements establishes the requirements that must be met to ensure the protection of people and the environment, both now and in the future. The requirements are governed by the objective and principles of the Safety Fundamentals. If the requirements are not met, measures must be taken to reach or restore the required level of safety. The format and style of the requirements facilitate their use for the establishment, in a harmonized manner, of a national regulatory framework. Requirements, including numbered 'overarching' requirements, are expressed as 'shall' statements. Many requirements are not addressed to a specific party, the implication being that the appropriate parties are responsible for fulfilling them.

Safety Guides

Safety Guides provide recommendations and guidance on how to comply with the safety requirements, indicating an international consensus that it

---

[1] See also publications issued in the IAEA Nuclear Security Series.

FIG. 1. The long term structure of the IAEA Safety Standards Series.

is necessary to take the measures recommended (or equivalent alternative measures). The Safety Guides present international good practices, and increasingly they reflect best practices, to help users striving to achieve high levels of safety. The recommendations provided in Safety Guides are expressed as 'should' statements.

APPLICATION OF THE IAEA SAFETY STANDARDS

The principal users of safety standards in IAEA Member States are regulatory bodies and other relevant national authorities. The IAEA safety standards are also used by co-sponsoring organizations and by many organizations that design, construct and operate nuclear facilities, as well as organizations involved in the use of radiation and radioactive sources.

The IAEA safety standards are applicable, as relevant, throughout the entire lifetime of all facilities and activities — existing and new — utilized for peaceful purposes and to protective actions to reduce existing radiation risks. They can be used by States as a reference for their national regulations in respect of facilities and activities.

The IAEA's Statute makes the safety standards binding on the IAEA in relation to its own operations and also on States in relation to IAEA assisted operations.

The IAEA safety standards also form the basis for the IAEA's safety review services, and they are used by the IAEA in support of competence building, including the development of educational curricula and training courses.

International conventions contain requirements similar to those in the IAEA safety standards and make them binding on contracting parties. The IAEA safety standards, supplemented by international conventions, industry standards and detailed national requirements, establish a consistent basis for protecting people and the environment. There will also be some special aspects of safety that need to be assessed at the national level. For example, many of the IAEA safety standards, in particular those addressing aspects of safety in planning or design, are intended to apply primarily to new facilities and activities. The requirements established in the IAEA safety standards might not be fully met at some existing facilities that were built to earlier standards. The way in which IAEA safety standards are to be applied to such facilities is a decision for individual States.

The scientific considerations underlying the IAEA safety standards provide an objective basis for decisions concerning safety; however, decision makers must also make informed judgements and must determine how best to balance the benefits of an action or an activity against the associated radiation risks and any other detrimental impacts to which it gives rise.

DEVELOPMENT PROCESS FOR THE IAEA SAFETY STANDARDS

The preparation and review of the safety standards involves the IAEA Secretariat and five Safety Standards Committees, for emergency preparedness and response (EPReSC) (as of 2016), nuclear safety (NUSSC), radiation safety (RASSC), the safety of radioactive waste (WASSC) and the safe transport of radioactive material (TRANSSC), and a Commission on Safety Standards (CSS) which oversees the IAEA safety standards programme (see Fig. 2).

All IAEA Member States may nominate experts for the Safety Standards Committees and may provide comments on draft standards. The membership of the Commission on Safety Standards is appointed by the Director General and includes senior governmental officials having responsibility for establishing national standards.

A management system has been established for the processes of planning, developing, reviewing, revising and establishing the IAEA safety standards.

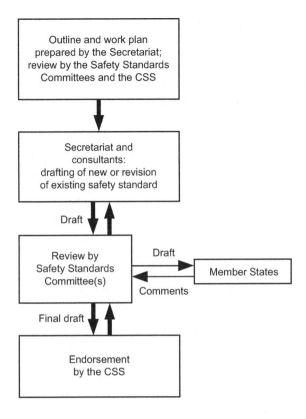

FIG. 2. The process for developing a new safety standard or revising an existing standard.

It articulates the mandate of the IAEA, the vision for the future application of the safety standards, policies and strategies, and corresponding functions and responsibilities.

INTERACTION WITH OTHER INTERNATIONAL ORGANIZATIONS

The findings of the United Nations Scientific Committee on the Effects of Atomic Radiation (UNSCEAR) and the recommendations of international expert bodies, notably the International Commission on Radiological Protection (ICRP), are taken into account in developing the IAEA safety standards. Some safety standards are developed in cooperation with other bodies in the United Nations system or other specialized agencies, including the Food and Agriculture Organization of the United Nations, the United Nations Environment Programme, the International Labour Organization, the OECD Nuclear Energy Agency, the Pan American Health Organization and the World Health Organization.

INTERPRETATION OF THE TEXT

Safety related terms are to be understood as defined in the IAEA Safety Glossary (see https://www.iaea.org/resources/safety-standards/safety-glossary). Otherwise, words are used with the spellings and meanings assigned to them in the latest edition of The Concise Oxford Dictionary. For Safety Guides, the English version of the text is the authoritative version.

The background and context of each standard in the IAEA Safety Standards Series and its objective, scope and structure are explained in Section 1, Introduction, of each publication.

Material for which there is no appropriate place in the body text (e.g. material that is subsidiary to or separate from the body text, is included in support of statements in the body text, or describes methods of calculation, procedures or limits and conditions) may be presented in appendices or annexes.

An appendix, if included, is considered to form an integral part of the safety standard. Material in an appendix has the same status as the body text, and the IAEA assumes authorship of it. Annexes and footnotes to the main text, if included, are used to provide practical examples or additional information or explanation. Annexes and footnotes are not integral parts of the main text. Annex material published by the IAEA is not necessarily issued under its authorship; material under other authorship may be presented in annexes to the safety standards. Extraneous material presented in annexes is excerpted and adapted as necessary to be generally useful.

CONTENTS

1. INTRODUCTION

BACKGROUND

1.1. Radioactive waste is, for legal and regulatory purposes, material for which no further use is foreseen that contains, or is contaminated with, radionuclides at activity concentrations greater than clearance levels as established by the regulatory body [1]. Radioactive waste must be managed safely and in such a way as to avoid imposing an undue burden on future generations; that is, the generations that produce radioactive waste have to seek and apply safe, practicable and environmentally acceptable solutions for its long term management, in accordance with IAEA Safety Standards Series No. SF-1, Fundamental Safety Principles [2].

1.2. Requirements for the management of radioactive waste are established in IAEA Safety Standards Series Nos GSR Part 5, Predisposal Management of Radioactive Waste [3], and SSR-5, Disposal of Radioactive Waste [4]. Management systems[1] for radioactive waste management[2] are subject to the requirements established in IAEA Safety Standards Series No. GSR Part 2, Leadership and Management for Safety [5].

1.3. This Safety Guide provides recommendations on meeting the requirements of GSR Part 2 [5] to provide confidence that the requirements for predisposal management of radioactive waste established in GSR Part 5 [3] and those for disposal of radioactive waste established in SSR-5 [4] will be met.

1.4. This Safety Guide supersedes IAEA Safety Standards Series Nos GS-G-3.3, The Management System for the Processing, Handling and Storage of

---

[1]   A management system is defined as a "set of interrelated or interacting elements (*system*) for establishing policies and objectives and enabling the objectives to be achieved in an efficient and effective manner" [1].

[2]   Radioactive waste management comprises all "administrative and operational *activities* involved in the handling, *pretreatment, treatment, conditioning, transport, storage* and *disposal* of *radioactive waste*" [1].

Radioactive Waste[3], and GS-G-3.4, The Management System for the Disposal of Radioactive Waste[4].

1.5. This Safety Guide identifies the need to consider nuclear security as well as safety; requirements and guidance on nuclear security are provided in the IAEA Nuclear Security Series publications.

1.6. The Joint Convention on the Safety of Spent Fuel Management and on the Safety of Radioactive Waste Management [6] and the supplementary Guidance on the Management of Disused Radioactive Sources issued as part of the Code of Conduct on the Safety and Security of Radioactive Sources [7] should be considered in developing management systems for the predisposal management and disposal of radioactive waste.

OBJECTIVE

1.7. The objective of this Safety Guide is to provide recommendations on developing and implementing management systems for safety during all steps of radioactive waste management — including processing (i.e. pretreatment, treatment and conditioning), storage and disposal, but excluding transport — and during related processes and activities as mentioned in para. 1.13. This Safety Guide also provides recommendations on effective leadership and culture for safety. The intention is that these recommendations will contribute to a high level of confidence that the following objectives will be achieved:

(a) Radioactive waste management activities will be conducted in compliance with the requirements.
(b) Radioactive waste packages will be of appropriate and consistent quality.
(c) The characteristics of radioactive waste packages will be sufficiently known.

---

[3] INTERNATIONAL ATOMIC ENERGY AGENCY, The Management System for the Processing, Handling and Storage of Radioactive Waste, IAEA Safety Standards Series No. GS-G-3.3, IAEA, Vienna (2008).

[4] INTERNATIONAL ATOMIC ENERGY AGENCY, The Management System for the Disposal of Radioactive Waste, IAEA Safety Standards Series No. GS-G-3.4, IAEA, Vienna (2008).

(d) Appropriate records will be kept that enable radioactive waste identification and decisions on whether radioactive waste packages and unpackaged waste[5] conform to the waste acceptance criteria for radioactive waste management facilities.

1.8. This Safety Guide is intended to be used by the regulatory body and organizations with responsibilities for directing, planning or undertaking the management of radioactive waste; it is also intended to be used by the suppliers[6] to such organizations of safety related services and products that support radioactive waste management. It will also be useful to members of the public and other interested parties.

SCOPE

1.9. This Safety Guide covers management systems for the following radioactive waste management activities:

(a) The minimization of radioactive waste generation;
(b) Processing, comprising pretreatment (e.g. collection, segregation, chemical adjustment, decontamination), treatment (e.g. volume reduction, removal of radionuclides from the waste, change of composition) and conditioning (e.g. immobilization, packaging, overpacking);
(c) Storage;
(d) Disposal (e.g. near surface disposal, geological disposal, borehole disposal).

1.10. This Safety Guide does not address the management system for the transport of radioactive waste, for which requirements are established in IAEA Safety Standards Series No. SSR-6 (Rev. 1), Regulations for the Safe Transport of Radioactive Material, 2018 Edition [8], and specific recommendations are provided in IAEA Safety Standards Series No. TS-G-1.4, The Management System for the Safe Transport of Radioactive Material [9].

---

[5] In this Safety Guide, the term 'waste' refers to radioactive waste unless otherwise stated.

[6] The supply chain, described as 'suppliers', typically includes designers, vendors, manufacturers and constructors, employers, contractors, subcontractors, and consigners and carriers who supply safety related items. The supply chain can also include other parts of the organization and parent organizations [5].

1.11. This Safety Guide covers management systems for the activities involved in managing all types of radioactive waste as described in IAEA Safety Standards Series No. GSG-1, Classification of Radioactive Waste [10], including the following:

(a) Activities that generate waste containing naturally occurring radionuclides;
(b) Activities in hospitals, laboratories and research and development facilities, and in industry;
(c) Decontamination of facilities or parts thereof;
(d) Decommissioning of facilities or parts thereof;
(e) Remediation (e.g. of areas affected by past activities);
(f) Activities to manage waste generated from incidents, including accidents, and from emergencies;
(g) Activities to manage legacy waste.

1.12. This Safety Guide provides guidance on the management system for the management of radioactive waste arising from remediation and from decommissioning, but not on any other aspect of decommissioning. Recommendations on the management system for decommissioning activities other than for the management of radioactive waste arising from decommissioning are provided in IAEA Safety Standards Series Nos. SSG-47, Decommissioning of Nuclear Power Plants, Research Reactors and Other Nuclear Fuel Cycle Facilities [11], and SSG-49, Decommissioning of Medical, Industrial and Research Facilities [12].

1.13. This Safety Guide also covers management systems for the following related processes and activities:

(a) Radioactive waste minimization;
(b) Radioactive waste characterization (e.g. to determine the radiological and physicochemical properties of waste);
(c) Clearance;
(d) Design and manufacture of radioactive waste containers and waste packages;
(e) Siting, design and construction of radioactive waste management facilities;
(f) Safety case development and safety assessment of radioactive waste management facilities;
(g) Authorization (e.g. licensing);
(h) Commissioning of radioactive waste management facilities;
(i) Operation of facilities for the predisposal management of radioactive waste;
(j) Operation of radioactive waste disposal facilities (e.g. activities, which can extend over several decades, involving receipt of radioactive waste, waste

emplacement in the disposal facility, backfilling and sealing, and any other operations in the period prior to closure);

(k) Closure of radioactive waste disposal facilities;

(l) Institutional control of radioactive waste disposal facilities, covering both active control (e.g. nuclear security, surveillance, monitoring) and passive control (e.g. preservation of records, restricted land use).

STRUCTURE

1.14. This Safety Guide addresses all the relevant requirements of GSR Part 2 [5], GSR Part 5 [3] and SSR-5 [4]. Section 2 identifies characteristics of radioactive waste management that influence leadership, management and culture for safety. Section 3 provides recommendations on responsibility for safety. Section 4 provides recommendations on leadership for safety. Section 5 provides recommendations on management for safety. Section 6 provides recommendations on culture for safety. Section 7 provides recommendations on the measurement, assessment and improvement of the management system. The Appendix identifies elements of the management system for radioactive waste management or its regulation.

2. CHARACTERISTICS OF RADIOACTIVE WASTE MANAGEMENT

2.1. Radioactive waste management has specific characteristics that affect leadership, management and culture for safety within organizations that have responsibilities for directing, planning, undertaking or regulating the management of radioactive waste, as well as within suppliers of safety related services and products within the supply chain. The following paragraphs identify such specific characteristics as a basis for the more detailed sections that follow.

2.2. Radioactive waste management is directed and undertaken by dedicated personnel in a range of organizations whose leadership needs to be able to foster the development of a strong culture for safety and establish and apply an effective management system. GSR Part 2 [5] emphasizes that leadership for safety, management for safety, a management system and a systemic approach (i.e. an approach relating to the system as a whole, in which the interactions between technical, human and organizational factors are duly considered) are essential to the establishment and implementation of adequate safety measures and the fostering of a strong culture for safety.

2.3. As stated in para. 1.5(b) of GSR Part 2 [5], management for safety includes the following:

"establishing and applying an effective management system. This management system has to integrate all elements of management so that requirements for safety are established and applied coherently with other requirements, including those for human performance, quality and security; and so that safety is not compromised by the need to meet other requirements or demands."

2.4. Radioactive waste management involves a wide range of activities, from simple, small scale, low hazard, repetitive tasks to complex, large scale, highly hazardous tasks at the limits of engineering capabilities. Because of the wide range of activities, it is important that the management system be developed and applied to a specific facility or activity using a graded approach (see Requirement 7 of GSR Part 2 [5]).

2.5. The following aspects warrant particular consideration in developing a management system for radioactive waste management:

(a) The provision of adequate human, financial and other resources for the safe management of radioactive waste. In accordance with the 'polluter pays' principle, an organization that generates radioactive waste is responsible for ensuring that funds are available for the waste to be managed properly.
(b) Ownership of and responsibility for radioactive waste. The responsibility for radioactive waste can change during its management. There should be clarity at all times regarding both ownership of the waste and responsibility for safety. In some jurisdictions, ownership of and responsibility for radioactive waste is transferred when the waste moves from one organization to another; in others, ownership of and responsibility for radioactive waste always remains with the original generator of the waste. While the owner should retain overall responsibility for the waste, the licensee of the facility where the waste resides is responsible for its safety.
(c) The possibility that national authorities or State organizations might need to take responsibility for radioactive waste because this responsibility cannot be discharged by the generator of the waste.
(d) The timescales involved in radioactive waste management can span many human generations (see para. 3.7 of SF-1 [2]); this raises issues for the

provision of resources, particularly the provision of financial and human resources, not only for technical and safety related aspects but also for knowledge management and for culture for safety.

(e) The need to ensure that radioactive waste generation is minimized (see para. 3.29 of SF-1 [2]) and that waste packages and unpackaged waste conform to the waste acceptance criteria of the receiving organization (see Requirement 12 of GSR Part 5 [3]).

(f) The views of interested parties on decisions on the management of radioactive waste.

(g) The selection of permanent options (i.e. discharge, clearance or disposal) for the release of radioactive waste from regulatory control.

(h) The need to ensure that, wherever possible, radioactive waste is in a passively safe condition.

(i) International best[7] practices and lessons from industry experience.

2.6. Radioactive waste is typically managed by a series of organizations (which might be privately or publicly owned) that carry out the sequence of required predisposal management and disposal steps. For example, radioactive waste generated by one organization may be transferred to another for processing (i.e. pretreatment, treatment and conditioning), to another for storage and to yet another for disposal (see Fig. 1). Figure 1 illustrates the systemic control of radioactive waste management by operating organizations at different facilities working under a series of management systems and under the governmental, legal and regulatory framework. Waste package specifications[8] and waste acceptance criteria facilitate the safe transfer of waste across the boundaries between management systems.

2.7. Processing, storage and disposal of radioactive waste may extend over a very long time (e.g. processing facilities and storage facilities for radioactive waste often operate for years or a few decades, and disposal facility operation may potentially last more than a hundred years). A radioactive waste management facility may therefore need to be managed over a long time period, potentially involving a series of different organizations with different leadership, cultures

---

[7] 'Best' as used here means the most effective in achieving a high general level of protection of people and the environment as a whole; further guidance on the meaning of best is included in IAEA Safety Standards Series No. GSG-9, Regulatory Control of Radioactive Discharges to the Environment [13].

[8] Waste package specifications comprise defined characteristics and properties of waste packages to provide for their acceptance at subsequent radioactive waste management facilities.

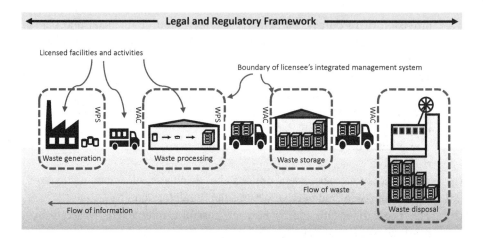

FIG. 1 Illustration of the systemic control of radioactive waste management by operating organizations at different facilities working under a series of management systems. WPS: waste package specifications; WAC: waste acceptance criteria.

and management systems, operating under successive authorities and national and international structures. The long duration of radioactive waste management has further implications, including the need to provide resources and to manage interdependencies between organizations and facilities over long time periods.

2.8. The long term nature of radioactive waste management, and particularly of radioactive waste disposal, also means that particular attention should be paid to the following:

(a) Maintaining the confidence of interested parties, including the public, that management supervision will be continuous over the period necessary;
(b) Establishing confidence that the long term performance of a radioactive waste disposal facility will meet regulatory requirements;
(c) Estimating costs and establishing the funding arrangements that will be necessary to continue to monitor and control the radioactive waste using the management system until active institutional control ceases;
(d) Ensuring continuity of understanding, knowledge, resources and safety culture over long time periods.

2.9. To gain the acceptance of the public and other interested parties for waste management operations, decision makers and leaders of the relevant organizations should place emphasis on the societal and ethical responsibilities to achieve safety now and in the future [14].

2.10. Without effective leadership, radioactive waste management programmes[9] have limited possibilities to succeed. Political leadership should create circumstances in which operating organizations can succeed safely. Examples in some States show that if political decisions are made at the right time, significant progress can be made in radioactive waste management. There should be good communication between and among decision makers and leaders of the relevant organizations involved in radioactive waste management, and a coordinated approach should be taken, particularly towards radioactive waste disposal (see e.g. Ref. [15]). Implementing disposal involves balancing several technical and sociopolitical requirements. Thus, it is important that all key actors — the government (at all levels), operating organizations and the regulatory body, plus supporting technical experts — be aware of the issues involved and facilitate the achievement of policy aims.

2.11. Leadership for safety in radioactive waste management, particularly in programmes for the disposal of radioactive waste, depends on senior management[10] possessing the following:

(a) Awareness and judgement;
(b) A clear view of long term policies and strategies for radioactive waste management, and the ability to communicate these effectively;
(c) The ability to distinguish strategically important issues;
(d) An understanding of which aspects are important to the safety of radioactive waste management;
(e) An understanding of the hazards and risks associated with the facilities and activities in their charge, and how these compare to other hazards and risks, so that they can oversee the development and application of the management system for radioactive waste management according to a graded approach;
(f) An understanding of which aspects of radioactive waste management are important to the public and other interested parties;
(g) The ability to explain and communicate the importance of safety and culture for safety in radioactive waste management.

---

[9] In a radioactive waste management programme, a group of related waste management projects is managed in a coordinated way and with a particular long term aim in order to obtain benefits and control not available from managing the projects individually.

[10] Senior management is defined as "The person or persons who direct, control and assess an organization at the highest level" [1].

3. RESPONSIBILITY FOR SAFETY

3.1. Requirement 1 of GSR Part 2 [5] states:

"The registrant or licensee — starting with the senior management — shall ensure that the fundamental safety objective of protecting people and the environment from harmful effects of ionizing radiation is achieved."

3.2. The prime responsibility for properly executing a particular radioactive waste management task (e.g. processing, storage or disposal, or a related activity such as the characterization of radioactive waste, clearance, or the design, construction, commissioning, operation and decommissioning or closure, as applicable, of a facility for predisposal management of radioactive waste or a radioactive waste disposal facility) rests with the operating organization[11] of the facility at which the waste management task is undertaken. The responsibilities of the regulatory body are defined in Requirement 3 of GSR Part 5 [3] and Requirement 2 of SSR-5 [4].

3.3. Paragraph 4.2 of GSR Part 2 [5] states that "Senior management shall be responsible for establishing safety policy." Senior management should define and implement an organization's safety policy based on the national policy and strategy for safety. The senior management of an organization responsible for a radioactive waste management facility or activity should be accountable for managing and demonstrating the safety of the facility or activity, consistent with the national policy and strategy for radioactive waste management and in compliance with regulatory requirements.

3.4. The senior management of an operating organization is responsible for developing goals, strategies, plans and objectives (see Requirement 4 of GSR Part 2 [5]), and for coordinating activities to achieve the fundamental

---

[11] The operating organization is defined as "Any organization or person applying for *authorization* or authorized to operate an *authorized facility* or to conduct an *authorized activity* and responsible for its *safety*....This includes, inter alia, private individuals, governmental bodies, *consignors* or *carriers*, *licensees*, hospitals and self-employed persons" [1]. Operating organization is synonymous with 'operator'. The licensee is defined as "The holder of a current *licence*. The *licensee* is the *person or organization* having overall responsibility for a *facility* or *activity*" [1]. The operating organization might not be the holder of the licence (e.g. the operator could be a supply chain organization). In practice, for an authorized facility, the operating organization is normally also the registrant or licensee. However, the separate terms are retained to refer to the two different capacities [1].

safety objective without unduly limiting the operation of facilities or the conduct of activities that give rise to radiation risks. Safety should be considered in all business decisions, activities and associated management system documentation, and protection must be optimized to provide the highest level of safety that can reasonably be achieved (see Principle 5 of SF-1 [2]). Paragraph 4.5 of GSR Part 2 [5] states that "Senior management shall ensure that goals, strategies and plans are periodically reviewed against the safety objectives, and that actions are taken where necessary to address any deviations".

3.5. Senior management should ensure that each step of radioactive waste management has consistent goals and objectives in order not to compromise the safety of the subsequent steps in the waste management process.

3.6. Senior management should prepare plans to ensure that essential functions can operate safely for a sustained period with significant employee absence (e.g. due to an influenza outbreak). The regulatory body should regularly review those plans.

3.7. Demonstrating safety involves the development of a safety case. Recommendations on the development of the safety case are provided in IAEA Safety Standards Series Nos GSG-3, The Safety Case and Safety Assessment for the Predisposal Management of Radioactive Waste [16], and SSG-23, The Safety Case and Safety Assessment for the Disposal of Radioactive Waste [17].

3.8. The senior management of an organization that manages radioactive waste is required to provide adequate resources to ensure that the organization manages the radioactive waste safely (see Requirement 9 of GSR Part 2 [5]). The senior management of such an organization should include in the management system provisions to deal with funding challenges, such as cost increases over time, cost uncertainties and risks, the availability of public and private funds, and unplanned events.

3.9. The clear allocation of accountabilities and responsibilities is essential to ensure safety throughout the management of radioactive waste. Senior management should ensure that it is clear within the management system when, how and by whom decisions are to be made and that appropriate records of decisions are documented (see para. 5.64).

3.10. Paragraph 3.3 of GSR Part 5 [3] states:

"It is possible that the predisposal management of radioactive waste will involve the transfer of radioactive waste from one operator to another, or that radioactive waste may even be processed in another State. In such situations, continuity of responsibility for safety is necessary throughout."

3.11. Appropriate management arrangements should be put in place for the transfer of waste to ensure that it is clear where responsibility lies and the exact point at which the transfer of responsibility takes place. These management arrangements should also include provision for the transfer of appropriate records and knowledge. The responsible body at any given time is required to have an adequate management system that meets Requirements 6–8 of GSR Part 2 [5] to ensure that safety is not compromised.

3.12. If an authorization for a radioactive waste management facility or activity is terminated at any time, the government should ensure that it is clear which parties are responsible for the safety of both the facility and the waste. As stated in footnote 5 of SF-1 [2], "Not having an authorization would not exonerate the person or organization responsible for the facility or activity from the responsibility for safety." If management and control of a site is necessary following termination of an authorization, the government should ensure that the necessary management and control is provided. For example, management arrangements may be necessary for monitoring purposes and for ensuring safety and security.

3.13. In some instances — for example, following closure of a disposal facility or the end of active institutional control of such a facility — responsibility might be transferred to the government. In such instances, the government should take over responsibility for record keeping, knowledge management and other passive institutional control measures, such as restricted land use.

3.14. There may be occasions on which there is radioactive waste for which no owner can be identified (e.g. if an orphan source is discovered and it is declared radioactive waste). In such cases, the government has to provide for the control of such waste (see para. 3.9 of SF-1 [2]). The government should arrange for the necessary resources and should assign clear responsibilities to appropriate organizations (e.g. the regulatory body) for the safe management of such waste. Organizations assigned responsibility for radioactive waste management should provide strong and effective leadership.

3.15. The senior management of a licensee that generates radioactive waste should liaise with the regulatory body and the operating organizations of relevant radioactive waste management facilities (including suitably qualified and experienced contractors, where relevant) with the aim of defining and optimizing arrangements for transfers of radioactive waste. Recommendations on the management of the supply chain to ensure that the waste can be safely managed through all steps of radioactive waste management are provided in Section 5.

3.16. The regulatory body is required to have legal authority to require the operating organization to provide the regulatory body with all necessary safety related information (see para. 2.13 of IAEA Safety Standards Series No. GSR Part 1 (Rev. 1), Governmental, Legal and Regulatory Framework for Safety [18]). The operating organization should initiate interactions with the regulatory body concerning the supply of necessary safety related information as soon as possible and before the processing of waste.

3.17. Senior management should direct and oversee the development, implementation, review and revision of emergency plans in accordance with para. 3.13 of GSR Part 5 [3]. Training, drills and exercises are also required to be provided for personnel relevant for emergency response, in accordance with Requirement 25 of IAEA Safety Standards Series No. GSR Part 7, Preparedness and Response for a Nuclear or Radiological Emergency [19].

3.18. Requirement 4 of GSR Part 7 [19] states that "**The government shall ensure that a hazard assessment is performed to provide a basis for a graded approach in preparedness and response for a nuclear or radiological emergency.**" The hazard assessment should take into account the characteristics of the waste, of the waste management facility, and of the site and its vicinity at each stage in the lifetime of the facility. Recommendations relevant to the management of large volumes of radioactive waste generated during a nuclear or radiological emergency are provided in IAEA Safety Standards Series No. GSG-11, Arrangements for the Termination of a Nuclear or Radiological Emergency [20], and further information is provided in Ref. [21].

4. LEADERSHIP FOR SAFETY

4.1. Requirement 2 of GSR Part 2 [5] states that "**Managers shall demonstrate leadership for safety and commitment to safety.**"

4.2. Senior management should recognize that the attitude of individuals and the culture for safety within an organization are influenced through leadership. To improve the culture for safety and help individuals to develop professionally, managers at all levels are required to demonstrate their commitment to safety as an overriding priority in resource allocation, in business planning, in documentation and in all waste management activities (see paras 3.1 and 3.2 of GSR Part 2 [5]).

4.3. Senior management should demonstrate a proactive and long term approach to safety issues in decision making in radioactive waste management.

4.4. Senior management should ensure that processes and procedures are incorporated in the management system to identify and manage human, technological and organizational factors affecting safety. This is particularly important for radioactive waste management programmes that involve numerous organizations in implementing technological solutions and carrying out radioactive waste management activities.

4.5. Senior management should promote and exercise open and effective communication on safety at all levels in the organization. Senior management should frequently and consistently share information concerning radioactive waste management with relevant personnel. Information with a bearing on safety, and on societal and economic elements, should be communicated, as appropriate, to personnel and other relevant interested parties, including the regulatory body. This is particularly important for radioactive waste management programmes in which various other parties and operating organizations are involved in the radioactive waste management process. Senior management should regularly seek feedback on how effective the leadership is in ensuring safety and improving the management system, and should ensure that necessary corrective actions are taken.

4.6. Where radioactive waste management is likely to occur over long periods, senior management should take particular care to ensure effective knowledge transfer (e.g. recording and archiving of information) and succession planning for continuing good leadership.

4.7. Senior management is required to establish expectations for behaviour with regard to safety (see para. 3.1(c) of GSR Part 2 [5]). Senior management should communicate to all personnel the expectations for the performance of safety related tasks by individuals and by teams. Managers should be familiar with the special characteristics of radioactive waste management, such as the need to manage interdependencies, and should communicate these characteristics to personnel.

4.8. Managers should, through their own actions, promote safe ways of working, be visibly involved in safety related activities and reinforce good practices. Contributing to the development and use of international safety standards is an example of a safety related activity that should promote safe ways of working. Managers are required to promote the values of the organization and encourage open, transparent and questioning behaviours (see para. 3.2 of GSR Part 2 [5]). Managers should also be able to recognize deteriorations in safety performance or safety related attitudes and should take immediate actions to respond to the situation.

4.9. Managers should promote ways for all personnel involved in radioactive waste management to participate in the development, implementation and continuous improvement of the management system (see Section 7), with the aims of optimizing protection and safety, and achieving the organization's safety goals. Where relevant, and taking account of the need to apply a graded approach, it should be possible for other parties affected by the waste management facility or activities (e.g. the public, waste generators, organizations involved in other parts of the waste management process, sub-contractors) to contribute to improvements to the management system. Participation in activities such as peer reviews and international research and development programmes on radioactive waste management can help personnel involved in radioactive waste management to gain a better understanding of the adequacy of the management system. Managers should motivate staff to share their perceptions on the adequacy of the management system, so that the organization can enhance safety performance.

4.10. Managers at all levels in the organization should actively promote the adoption of effective measures to respond to events (including near misses) and to learn lessons from operating experiences and from identifying good practices (see para. 6.7 of GSR Part 2 [5]).

4.11. Managers should also have administrative, communication and interpersonal skills. Managers should develop their skills and support their colleagues and staff, both internally and in other organizations involved in radioactive waste management, to systematically develop their skills, solve problems and resolve conflicts.

4.12. Paragraph 3.3 of GSR Part 2 [5] states:

"Managers at all levels in the organization:

(a) Shall encourage and support all individuals in achieving safety goals and performing their tasks safely;

(b)   Shall engage all individuals in enhancing safety performance;

(c)   Shall communicate clearly the basis for decisions relevant to safety."

4.13. Managers should communicate routinely and often with individuals working in the organization. Managers should check that individuals understand their safety goals and how they are expected to perform their tasks safely. Managers should observe individuals' work, monitor safety performance and provide feedback on the performance of safety related tasks. Good performance should be recognized and, as appropriate, praised and rewarded. Managers should motivate individuals and assist them in maintaining and increasing their self-esteem and pride in work performed; this is particularly relevant for radioactive waste management tasks, which need to be performed to produce consistent, high quality waste products, sometimes over long timescales. Managers should engage with all the individuals working in the organization to identify areas of weaker performance and devise appropriate solutions for enhancing safety performance. The implementation of improvements and enhancements should be facilitated and encouraged. Training and coaching in radioactive waste management tasks should be provided as appropriate. Managers are required to make clear the basis for safety related decisions; this involves providing rational explanations for decisions, including information on their understanding of what is important to the safety of the radioactive waste management facilities and activities, and on other relevant factors, supported by appropriate data and evidence.

5. MANAGEMENT FOR SAFETY

RESPONSIBILITY FOR THE INTEGRATION OF SAFETY INTO THE MANAGEMENT SYSTEM

Responsibility of senior management for the management system

5.1.   Requirement 3 of GSR Part 2 [5] states that **"Senior management shall be responsible for establishing, applying, sustaining and continuously improving a management system to ensure safety."** Senior management remains responsible for the management system even when an external organization is involved in its development or improvement.

5.2. The development of a management system for an organization should take into account, as appropriate, the following:

(a) International standards such as ISO 9001:2015 for quality management systems [22], ISO 14001:2015 for environmental management systems [23] and ISO 45001:2018 for occupational health and safety management systems [24];
(b) The national legal framework, and regulatory requirements and guidance;
(c) Best practices in the nuclear and radioactive waste management industries;
(d) The organization's responsibilities, short term and long term objectives, and strategic plans.

5.3. Irrespective of the codes, standards and requirements used in developing the management system, the design of the management system should incorporate processes and procedures both to comply with these codes, standards and requirements and to demonstrate compliance.

5.4. Senior management should ensure that all radioactive waste management activities are undertaken in compliance with the management system. Senior management should ensure that the management system continues to be properly implemented, assessed and improved, especially during periods of change, and that relevant personnel are informed of any changes and the reason for their introduction and are trained in the new processes and procedures.

5.5. Senior management should put in place arrangements to ensure that managers at all levels demonstrate commitment to the establishment, implementation, assessment and continuous improvement of the management system. Senior management should ensure that, where appropriate, the management system is capable of dealing with long term aspects, such as changes in responsibilities, as well as any interdependencies among waste generators, predisposal management facilities and activities, and waste disposal facilities and activities.

5.6. Senior management should recognize that radioactive waste management may be affected by many factors. In particular, senior management should recognize that radioactive waste disposal involves facilities built in the natural environment, which will need to perform safely over a long period. National, regional and international policies and principles for radioactive waste management, including radioactive waste disposal and industry standards for management systems, will evolve over the extended period of time for which radioactive waste management activities may continue. In addition, policy decisions (e.g. regarding the reprocessing of spent fuel) and technological innovations and advances may lead to changes in the overall radioactive waste management strategy. Irrespective of all these factors, senior management retains responsibility for the safety of facilities and activities at all times. Senior management should, therefore, demonstrate continuous commitment to developing, implementing and improving the management system as a prerequisite to ensuring and maintaining safety.

5.7. Senior management should appoint individuals from within the organization to have specific responsibilities and authorities for the management system in the following areas[12]:

(a) Coordinating the development and implementation of the management system and its assessment and continuous improvement;
(b) Measuring and reporting on the performance of the management system, including its influence on safety and safety culture, and any need for improvement;
(c) Resolving any potential conflicts between requirements on radioactive waste management and requirements on other fields of activity, such as mining safety and environmental protection (e.g. groundwater protection), and between different elements and processes of the management system.

5.8. Individuals should not be given overlapping or conflicting responsibilities and authorities.

5.9. Senior management is required to retain accountability for the management system even where other individuals are assigned responsibility for coordinating the development, application and maintenance of the management system (see para. 4.1 of GSR Part 2 [5]). Senior management should appoint an individual manager to have overall responsibility for the organization's management system that applies to the radioactive waste management programme.[12] Furthermore, senior management should ensure, when defining that person's duties, that all the waste management activities are covered in a comprehensive and coherent manner, and that these activities are covered continuously over an appropriate period of time. This is especially important for radioactive waste storage facilities and for radioactive waste disposal facilities where there could be responsibilities that extend for long periods of time.

---

[12] Senior management may perform some or all of these roles themselves (e.g. in small organizations).

5.10. For each process within the management system (see paras 5.87–5.117), senior management should ensure that a designated individual[12] is given the authority and responsibility for the following:

(a) Developing and documenting the process and maintaining the necessary supporting documentation;
(b) Ensuring that there is effective interaction between process interfaces;
(c) Ensuring that process documentation is consistent and appropriate for the waste management facilities and activities;
(d) Ensuring that the records necessary for demonstrating that the process results have been achieved are specified in the process documentation;
(e) Monitoring and reporting on the performance of the process;
(f) Promoting improvement in the process;
(g) Ensuring that the process and any changes subsequently adopted are aligned with the goals, strategies, plans and objectives of the organization and are consistent with the organization's safety policy.

5.11. The roles and responsibilities for safety in radioactive waste management may continue for a long time and may change during this time. Responsibilities for radioactive waste may transfer between organizations and may even transfer between States (e.g. in accordance with agreements on the repatriation of waste). Management systems should be designed to ensure continuity in managing facilities and activities, and should contain provisions for managing changes, for example, in the following:

(a) The ownership of radioactive waste and radioactive waste management facilities;
(b) Management arrangements;
(c) The regulatory body;
(d) National and international legislation and standards;
(e) Land use policies in relation to the institutional control of facilities.

5.12. When the management arrangements for radioactive waste management facilities are changed (e.g. if public organizations are privatized, if new organizations are created, if existing organizations are combined or restructured, if responsibilities are transferred between organizations, if an operating organization undergoes internal reorganization of its management structure or reallocation of resources), the possible need to adapt the management system should be assessed by senior management, while ensuring that the management system continues to be properly implemented, assessed and improved.

Goals, strategies, plans and objectives

5.13. Requirement 4 of GSR Part 2 [5] states that "**Senior management shall establish goals, strategies, plans and objectives for the organization that are consistent with the organization's safety policy.**"

5.14. The goals, strategies, plans and objectives that are established by senior management should be derived and documented in accordance with the management system and should recognize that both short term and long term safety aspects are involved in radioactive waste management. The goals, strategies, plans and objectives should give paramount importance to safety and should seek to adhere to the waste hierarchy (see para. 3.29 of SF-1 [2] and Ref. [25]).

5.15. The goals, strategies, plans and objectives should include appropriate means of considering the concerns and expectations of interested parties in decision making (see paras 5.21–5.31), and should be communicated effectively and consulted upon, as appropriate.

5.16. Radioactive waste management strategies should be developed taking full advantage of opportunities and synergies arising from national, regional and international cooperation and experience, where appropriate. Radioactive waste management strategies should include milestones and clear time frames for the achievement of these milestones.

5.17. Senior management is also responsible for establishing the safety policy of the organization (see para. 4.2 of GSR Part 2 [5]) and should ensure that this policy is documented in the management system. Senior management should also ensure that the management system is updated if goals, strategies, plans, policies or objectives are changed. Hence, the management system documentation will consist of a dynamic collection of living documents.

5.18. The safety policy should include the following:

(a)   A statement that safety will be given overriding priority, in ensuring that other relevant requirements and provisions (e.g. for nuclear security) are also met.
(b)   A statement that the safety policy will comply with applicable national, regional and international policies, strategies and regulations on radioactive waste management.
(c)   A statement that the safety policy will take into account the views, attitudes, concerns and expectations of the public and other interested parties, in

relation to safety (and, where applicable, in relation to restrictions on the use of land and natural resources).

(d) A statement that the safety policy will be appropriate to the objectives and the activities of the organization.

(e) Statements on how societal and economic considerations are taken into account with regard to safety.

(f) A commitment to comply with the management system and to seek continuous improvement.

(g) A commitment to support the development of a strong culture for safety.

(h) An appropriate framework for action and for establishing and reviewing goals and objectives at all levels. Where possible, safety goals and objectives should be measurable.

(i) A commitment to periodic review to ensure continuing suitability and applicability.

(j) A mechanism for the safety policy to be effectively communicated, understood and followed within the organization.

(k) A commitment to minimizing the generation of radioactive waste, to safe storage and to safely disposing in a timely manner of radioactive waste that cannot otherwise be removed from regulatory control.

5.19. The management system for each organization carrying out work to implement, support, regulate or evaluate a radioactive waste management programme should include a process for the periodic review of the organization's safety policy. Such reviews should, as appropriate, take into account the following:

(a) Changes in the legal and regulatory framework for radioactive waste management;

(b) Changes in regulatory requirements for radioactive waste management;

(c) International developments (e.g. new standards, conventions, agreements on information exchange);

(d) Technological advances;

(e) Learning from operating experience and from events;

(f) Non-conformances and corrective measures and preventive measures, and the results of safety assessments;

(g) Results of national, regional and international reviews and assessments of radioactive waste management programmes and developments in radioactive waste management practices;

(h) Results of internal and external audits, peer reviews and inspections of waste management facilities and activities;

(i) Results of environmental monitoring and other types of monitoring and surveillance.

5.20. All personnel within the organization should understand the safety policy and should accept personal accountability for their own conduct in meeting its objectives (see para. 5.2(b) of GSR Part 2 [5]).

Interaction with interested parties

5.21. Requirement 5 of GSR Part 2 [5] states that "**Senior management shall ensure that appropriate interaction with interested parties takes place.**"

5.22. GSR Part 2 [5] also states:

"4.6. Senior management shall identify interested parties for their organization and shall define an appropriate strategy for interaction with them.

"4.7. Senior management shall ensure that the processes and plans resulting from the strategy for interaction with interested parties include:

(a) Appropriate means of communicating routinely and effectively with and informing interested parties with regard to radiation risks associated with the operation of facilities and the conduct of activities;

(b) Appropriate means of timely and effective communication with interested parties in circumstances that have changed or that were unanticipated;

(c) Appropriate means of dissemination to interested parties of necessary information relevant to safety;

(d) Appropriate means of considering in decision making processes the concerns and expectations of interested parties in relation to safety."

5.23. Senior management should ensure that the management system includes an appropriate process for the identification of interested parties. Different interested parties are likely to have different needs and viewpoints. Therefore, it is important to identify the interested parties and to determine their particular interests, needs, expectations and concerns. This is essential for selecting effective approaches to communication, information dissemination, consultation and decision making.

5.24. The process for the identification of interested parties should take account of specific national, regional and local attributes. Experience has shown that it is not always straightforward to define, for example, potential host communities, or local or potentially affected communities, or relevant regional bodies or organizations (see e.g. Ref. [26]). The characteristics of the radioactive waste management facilities and activities, and their associated hazards and risks, should also be considered when identifying interested parties and adopting and implementing methods for interaction. The process for the identification of interested parties should be flexible and able to cope with changing circumstances and the possible emergence of different interested parties over time. This will be particularly relevant for radioactive waste management facilities that operate over long periods.

5.25. Gaining societal acceptance for radioactive waste disposal facilities can be particularly difficult. Some programmes for the siting and development of radioactive waste disposal facilities have therefore adopted approaches involving partnerships with local communities. Partnership approaches involve collaborative working relationships between communities and the operating organization. The key feature of the partnership approach is the empowerment of local communities in decisions that affect their future. Such partnership approaches may include seeking volunteer communities. A volunteer community is one that has expressed interest in participating in a process to determine the suitability of a site for a radioactive waste management facility. Such an expression of interest may be conveyed by appropriate representatives of the community (e.g. from a local governing body) and may be made in response to an invitation by the operating organization or by the government, or may be an unsolicited offer. A volunteer community should have either a formal or informal right to withdraw from the process and may receive an appropriate community benefits package. The process to determine the approach to be followed is often defined by the government. The senior management of the operating organization should be involved in the process to define the approach to be followed. The management system for a radioactive waste disposal programme should include specific processes for participating in the approach to identifying and interacting with interested parties.

5.26. IAEA Safety Standards Series No. GSG-6, Communication and Consultation with Interested Parties by the Regulatory Body [27], describes the roles of typical interested parties, including employees, the public, news and social media, local liaison groups (or committees), special interest groups, governmental authorities and decision makers, professional bodies, international organizations and national regulatory bodies. In addition to these, interested parties may also include operating organizations, funding entities and trade unions, land owners, and industry, more generally, and contractors, as well as organizations involved in emergency preparedness and response. The management system should include processes to ensure that personnel, particularly those interacting with interested parties outside the organization, are appropriately informed of decisions and activities of the organization, and of other relevant safety related information. Personnel should be aware that their communications might affect how the organization is perceived, particularly if media channels that can reach large audiences are used for such communications (e.g. statements to journalists, comments on web sites or social media).

5.27. The expectations of interested parties should be taken into account when developing the management system for radioactive waste management. Aspects that might need to be considered when developing the management system include the following:

(a) Legal aspects (e.g. the governmental, legal and regulatory framework and regulatory requirements relating to topics such as non-radiation-related safety, environmental protection, mining).
(b) Restrictions on the transport of radioactive material and hazardous materials across borders and boundaries between different areas.
(c) Nuclear security provisions that may be necessary, as appropriate, for nuclear material and other radioactive material.
(d) Operational limitations, including those derived from agreements with the responsible national, regional and local authorities or organizations, or arising from operating logistics. For example, it may not be possible in practice to operate to the capacity authorized; voluntary agreements may have been reached with interested parties other than the regulatory body that have the effect of limiting operations in a way that goes beyond, or is different from, the limits and conditions contained in the authorization (e.g. stacking radioactive waste so that local residents cannot see it, running operations at only certain times of the day, organizing transport to avoid certain routes); logistical reasons might exist for not proceeding with certain waste management steps or not proceeding with them as fast as might be desired (e.g. the next facility in the waste management chain may not be ready to receive conditioned waste into its stores).
(e) The needs, expectations and concerns of other organizations in the waste management chain.
(f) Public attitudes, concerns and expectations about safety in relation to radioactive waste management (e.g. concerns about the consequences of discharges, the adequacy and reliability of long term organizational and financial arrangements, site selection and site characterization processes, the degree of confidence in safety during operations and in the long term, and the ability to respond to problems that might arise).
(g) Public concerns and cultural expectations relating to restrictions on the use of land (e.g. historically significant sites, sacred sites) and resources (e.g. minerals, oil and gas, water).

(h)   Other concerns of interested parties (e.g. cultural expectations about working hours and the composition of the workforce, societal expectations regarding the distribution of risks and benefits, political choices about activities and sustainable development).

5.28. The operating organization should ensure that all necessary arrangements are put in place for informing the public and other interested parties about potential impacts (e.g. radiation risks) associated with radioactive waste management facilities and activities. Information should be provided prior to starting activities and thereafter during the conduct of the activities. The process for interacting with interested parties should include methods and procedures aimed at resolving any conflicts that arise.

5.29. Processes and procedures for communicating and interacting with interested parties should be designed to be suitable for the long periods of time potentially involved in radioactive waste management.

5.30. Communication with interested parties should include information on aspects such as the following:

(a)   The safety case for radioactive waste management facilities and activities, the status of operations, and plans for the future;
(b)   The occurrence of any incidents, including accidents, the measures taken to respond to them and the actions taken to prevent a recurrence;
(c)   The safety, societal and economic impacts of the radioactive waste management activities;
(d)   Changes in management arrangements and the continuity of responsible management;
(e)   Maintenance of adequate financial resources to support the radioactive waste management activities;
(f)   Opportunities for, and results from, the involvement of interested parties in decision making;
(g)   Responses to questions and concerns from interested parties.

5.31. Except for circumstances relating to security or commercial confidentiality, open communication should be promoted and exercised within all levels in the organization and with the public and other interested parties. The aim should be to work closely with interested parties throughout the lifetime of a radioactive waste management facility in order to build relationships, foster understanding of issues and facilitate more inclusive decision making and the resolution of issues.

Interdependencies in radioactive waste management

5.32. Requirement 6 of GSR Part 5 [3] states: "**Interdependences among all steps in the predisposal management of radioactive waste, as well as the impact of the anticipated disposal option, shall be appropriately taken into account.**"

5.33. The management system should include processes and procedures to take into account interdependencies between the steps in the minimization of radioactive waste generation and radioactive waste management (i.e. handling, pretreatment, treatment, conditioning, transport, storage and disposal).

5.34. In order to address the interdependencies between the steps in radioactive waste management, possible impacts on subsequent waste management steps should be identified and assessed, and appropriate decisions should be made concerning the choice of option for the current step. For example, the option selected for processing a particular radioactive waste stream should produce waste packages that are suitable for subsequent storage and disposal steps. The assessments of possible impacts of interdependencies, and the decisions made concerning the choice of option for the current waste management step and the reasons for them, should be documented.

5.35. Making these assessments and decisions will require coordination and the timely exchange of information between the organizations involved. For example, purchase details for sealed radioactive sources (e.g. manufacturer, radionuclide content, initial activity) should be preserved, together with a history of the usage of each source and records of any instances of damage, and this information should be made available to organizations in the waste management chain concerned with the processing, storage and disposal of sealed radioactive sources that have become disused and been declared as radioactive waste.

5.36. With the possible exception of situations during a nuclear or radiological emergency in which large volumes of radioactive waste are generated (see GSG-11 [20] and Ref. [21]), radioactive waste should not be handled, treated, conditioned or stored in a manner that would make the waste more difficult to manage at a later stage in the waste management process.

5.37. The development and use of waste package specifications and waste acceptance criteria is one of the main methods used to take account of interdependencies in the radioactive waste management process. An example of such arrangements is described in Ref. [28]. Within a waste management programme, standardization (e.g. standard waste containers, standard storage arrangements) can also be helpful in managing interdependencies.

5.38. The processes and procedures included in the management system for addressing and managing interdependencies should enable the safety and effectiveness of the radioactive waste management steps to be considered in an integrated manner. Paragraph 3.22 of GSR Part 5 [3] states:

"This includes taking into account the identification of waste streams, the characterization of waste, and the implications of transporting and disposing of waste. There are two issues in particular to be addressed: compatibility (i.e. taking actions that facilitate other steps and avoiding taking decisions in one step that detrimentally affect the options available in another step) and optimization (i.e. assessing the overall options for waste management with all the interdependences taken into account). The use of well managed information of good quality is key to both aspects."

5.39. A key feature of the radioactive waste management process shown in Fig. 1 is that information flows in both directions. The flow of information from disposal towards storage, processing and the minimization of waste generation should be used in the design and optimization of the steps earlier in the process so that they result in waste forms, waste containers and waste packages that are suited to the subsequent waste management steps. Some States have implemented programmes and projects aimed at such optimization; examples are described in Refs [29, 30]. As well as addressing interdependencies, these programmes should aim to optimize the radioactive waste management process as a whole.

THE MANAGEMENT SYSTEM

Integration of the management system

5.40. Requirement 6 of GSR Part 2 [5] states that "**The management system shall integrate its elements, including safety, health, environmental, security, quality, human-and-organizational-factor, societal and economic elements, so that safety is not compromised.**"

5.41. The management system is required to be aligned with the safety goals of the organization (see para. 4.8 of GSR Part 2 [5]).

5.42. Paragraph 4.9 of GSR Part 2 [5] states:

"The management system shall be applied to achieve goals safely, to enhance safety...by:

(a) Bringing together in a coherent manner all the necessary elements for safely managing the organization and its activities;

(b) Describing the arrangements made for management of the organization and its activities;

(c) Describing the planned and systematic actions necessary to provide confidence that all requirements are met;

(d) Ensuring that safety is taken into account in decision making and is not compromised by any decisions taken."

5.43. The management system is required to specify clearly the organizational structures, processes, responsibilities, accountabilities, levels of authority and interfaces within the organization and with external organizations (see para. 4.11 of GSR Part 2 [5]). The organizational structures should be clear and the reasons for the structures should be explained and justified (e.g. so that personnel working within the organization can understand the reasons for the structures and are better able to contribute to improving the management system). The management of processes and activities is addressed in paras 5.87–5.117. The identification of responsibilities is particularly important for radioactive waste management programmes in which the waste generator transfers responsibility for safety to a series of waste management operating organizations. The point at which responsibility changes should be clearly defined and documented within the management system, ensuring that safety is not compromised. The management systems established by each of these organizations should include contingency measures to deal with unexpected occurrences, such as accidents and cases where waste acceptance criteria are not met.

5.44. In integrating the elements of the management system, synergies should be identified to simplify compliance with different requirements and facilitate a consistent approach. This is particularly important for radioactive waste management programmes of a long term nature, because of the potential for responsibilities to change and the interdependencies between different waste management stages. Consequently, management systems need to be flexible and able to manage change as described in Section 7.

5.45. The management system should be developed so that it covers all activities to be carried out during radioactive waste management. Requirement 2 of GSR Part 1 (Rev. 1) [18] states that **"The government shall establish and maintain an**

appropriate governmental, legal and regulatory framework for safety within which responsibilities are clearly allocated." Importantly, at the national level, the governmental, legal and regulatory framework should ensure that the management systems of the various operating organizations interface well with one another. The effectiveness of the interfaces between the various management systems should be assessed and documented. The effectiveness of the governmental, legal and regulatory framework should be evaluated, for example in terms of assessing the competence of operating organizations, the interfaces between the management systems and operations, and the achievement of the national radioactive waste management strategy. This integration is necessary so that the interfaces between governmental arrangements and the management systems and operations of operating organizations, and between different management systems and operations, are appropriate and properly managed.

5.46. The management system should cover all steps and periods in the radioactive waste management process in an integrated manner, including the identification of measures to be taken during the period of institutional control after the closure of disposal facilities. The duration of the period of post-closure institutional control is required to be justified in the relevant safety case (see paras 4.23 and 4.24 of SSR-5 [4]).

5.47. In developing the management system, senior management should ensure that the overall strategy for the waste management programme is reflected in detailed processes and intended outputs, and in the criteria for the characteristics and properties of radioactive waste and waste packages requiring predisposal management and disposal.

5.48. The management system should be designed so that it can be adapted, as necessary, to accommodate future technological advances and changes in waste acceptance criteria that could have implications for the radioactive waste management steps leading to its final disposal.

5.49. The management system should include provisions to ensure that the development of detailed processes for waste management is informed by safety assessment, and that there is an iterative interaction between facility design and safety assessment. This includes the following design–assessment cycle:

(a) Tentative specifications for waste and for waste packages, and criteria for their acceptance, should be developed when the sequence of waste management activities is first conceived. Once site specific and facility specific information is available, detailed waste package specifications and waste acceptance criteria should be developed based on safety assessment results and other data, as appropriate.

(b)   The level of safety provided by various combinations of waste, waste packages and alternative facility designs should be assessed.
(c)   The feasibility of implementing the possible designs should be evaluated.
(d)   The effect on safety of any potential design improvements to radioactive waste management facilities should be assessed.

5.50. The management system should include a process and procedures that provide for the design–assessment cycle described in para. 5.49 to be repeated, usually several times. This process should result in a set of waste characteristics, facility design specifications, and safety assessments and safety cases, and these should be used to guide the development of the entire set of waste management activities.

5.51. When integrating the elements of the management system, long term aspects of the radioactive waste management programme should be considered, such as the following:

(a)   Providing adequate (e.g. human, infrastructure, financial) resources (e.g. for site maintenance), taking account of the amounts and types of waste to be managed and the storage and disposal options that have been adopted. The adequacy of resources should be reviewed periodically, particularly for facility development and operational periods that may extend over decades.
(b)   Preserving technology and knowledge, and transferring knowledge to people joining the programme in the future.
(c)   Retaining or transferring ownership of waste and waste management facilities.
(d)   Succession planning for managers and personnel.
(e)   Continuing arrangements for interacting with interested parties.

Application of the graded approach to the management system

5.52. Requirement 7 of GSR Part 2 [5] states that **"The management system shall be developed and applied using a graded approach."**

5.53. Paragraph 4.15 of GSR Part 2 [5] states:

"The criteria used to grade the development and application of the management system shall be documented in the management system. The following shall be taken into account:

(a)   The safety significance and complexity of the organization, operation of the facility or conduct of the activity;

(b)   The hazards and the magnitude of the potential impacts (risks) associated with the safety, health, environmental, security, quality and economic elements of each facility or activity…;

(c)   The possible consequences for safety if a failure or an unanticipated event occurs or if an activity is inadequately planned or improperly carried out."

5.54. The safety significance of the various facilities and activities within the radioactive waste management programme should be determined and documented. Resources should be allocated, and processes should be designed, to control these facilities and activities effectively and efficiently, in accordance with their safety significance. Controls will differ for different waste management facilities and activities, and should be applied in accordance with a graded approach.

5.55. Applying a graded approach means that the stringency of the control measures and conditions to be applied to a system are commensurate with the likelihood and possible consequences of, and the level of risk associated with, a loss of control [1]. The application of a graded approach is intended to guide the degree of control applied to a radioactive waste management process, facility or activity to ensure that the degree of control reflects the importance of the function of the process, facility or activity and the associated risk and to ensure appropriate use of resources. The application of a graded approach should not be used as a justification for not applying all the necessary management system elements and quality management controls, for not meeting regulatory requirements, or for performing less than adequate technical assessments of items that are less important to safety, or as a basis for inadequate practices. Following a graded approach is not a valid reason for not determining the adequacy of any activity affecting safety.

5.56. The method for applying the graded approach should be documented in the management system. Effective management involves the proportionate application of controls to facilities and activities on the basis of various criteria, including the following:

(a) The quantities of waste, the potential hazards associated with the waste, the necessary degree of isolation and the timescale over which isolation needs to function;
(b) The potential dispersibility of the waste, the potential mobility of the waste and the necessary degree of waste containment;
(c) The time period before disposal;
(d) Experience with, and the maturity of, the technology used in radioactive waste management activities;
(e) The reliability of equipment and its function in relation to safety;
(f) The complexity and degree of standardization of the waste management activities;
(g) The novelty and maturity of waste management activities, particularly for 'first of a kind' activities;
(h) The size of the operating organization, the number and complexity of interfaces with other organizations in the radioactive waste management process, and the organization's culture for safety;
(i) Public perception of radiation hazards and the risks associated with the radioactive waste;
(j) Government policy (e.g. on nuclear power generation and radioactive waste management);
(k) Possible future human activities and realistic exposure scenarios;
(l) External events and processes that could affect facilities, particularly long term events and processes such as ground settlement, erosion and climate change for facilities that will be operated for long periods;
(m) The likelihood of incidents, including accidents, and provisions for mitigating their consequences if they were to occur.

5.57. The graded approach to applying the management system should be based on the findings of appropriate assessment (e.g. safety assessment, hazard assessment). When applying the management system for radioactive waste management facilities, consideration should be given as appropriate to the following:

(a) The level of detail of work instructions and supporting documentation;
(b) The level of qualification and training of personnel;
(c) The quantity, level of detail and retention times of records;
(d) The level of detail and frequency of testing, surveillance and inspection;
(e) The equipment to be included in the configuration management at the facility;

(f)   Key performance indicators;

(g)   Equipment calibration;

(h)   The need to monitor the condition of equipment, waste and the facility;

(i)   The traceability of items, including radioactive waste;

(j)   The availability of storage, conditions of storage and control of associated records;

(k)   The level of reporting and the authority to act on non-conformances and to implement corrective actions;

(l)   The maturity of the safety assessments and safety cases and how well they represent the current state of the facilities and activities, and the requirements for periodic safety assessment;

(m)   The scope, frequency and detail of safety audits of radioactive waste management facilities and activities;

(n)   The scope and detail of any environmental monitoring programme.

5.58. Further information on the use of a graded approach in the application of the management system requirements for facilities and activities can be found in Ref. [31].

Documentation of the management system

5.59. Requirement 8 of GSR Part 2 [5] states that "**The management system shall be documented. The documentation of the management system shall be controlled, usable, readable, clearly identified and readily available at the point of use.**"

5.60. GSR Part 2 [5] also states:

"4.16. The documentation of the management system shall include as a minimum: policy statements of the organization on values and behavioural expectations; the fundamental safety objective; a description of the organization and its structure; a description of the responsibilities and accountabilities; the levels of authority, including all interactions of those managing, performing and assessing work and including all processes; a description of how the management system complies with regulatory requirements that apply to the organization; and a description of the interactions with external organizations and with interested parties.

"4.17. Documents shall be controlled. All individuals responsible for preparing, reviewing, revising and approving documents shall be competent to perform the tasks and shall be given access to appropriate information on which to base their input or decisions.

"4.18. Revisions to documents shall be controlled, reviewed and recorded. Revised documents shall be subject to the same level of approval as the initial documents.

"4.19. Records shall be specified in the management system and shall be controlled. All records shall be readable, complete, identifiable and easily retrievable.

"4.20. Retention times of records and associated test materials and specimens shall be established to be consistent with the statutory requirements and with the obligations for knowledge management of the organization. The media used for records shall be such as to ensure that the records are readable for the duration of the retention times specified for each record."

Documenting the management system

5.61. The documentation of the management system should be developed to be understandable, unambiguous and user friendly using a hierarchical approach where appropriate. A controlled document is a document that is approved and maintained. Controlled documents should be readable, complete, readily identifiable and easily available at the point of use. Controlled documents should be signed and dated and should bear a reference including the revision state. The number of pages and annexes in a controlled document should be clearly shown. Changes between revisions of controlled documents should be clearly marked. Further guidance on the document control process can be found in IAEA Safety Standards Series No. GS-G-3.1, Application of the Management System for Facilities and Activities [32].

5.62. Policies (i.e. statements of goals and objectives), strategies, plans, safety cases, safety assessments, management system processes and procedures, instructions, specifications and drawings (or representations in other media), training materials and any other documents that describe radioactive waste management processes and activities, specify requirements or establish waste package specifications should be controlled. It should be ensured that document users are aware of, and use, appropriate and correct documents.

5.63. Radioactive waste management activities differ greatly in size and complexity, may involve a number of organizations and may continue over extended periods of time[13], during which management practices and operating processes can evolve significantly. Particular attention should be paid to ensuring that the documents used to control work processes remain relevant, current and understandable, and are available to the organizations involved at the locations and times at which they are needed.

Record keeping and management

5.64. In addition to documenting the management system, traceable records should be created that describe and characterize the radioactive waste and the waste management activities undertaken. The records should include various types of information, including the following, as appropriate:

(a) The origin of the waste and the processes by which it was generated;
(b) The physical and chemical forms and properties of the waste (e.g. of the materials used in waste conditioning and their radionuclide retention properties);
(c) The activity concentration and total activity of radionuclides in the waste;
(d) The mass, activity concentration and total activity of fissile nuclides in the waste;
(e) The type of waste package;
(f) The radiation level at the surface of the waste package;
(g) The level of surface contamination on the waste package;
(h) The mass and weight of the waste or waste package;
(i) The dates of waste processing;
(j) The methods and instruments used to describe and characterize the waste.

5.65. The range of information and the level of detail to be recorded should be specified in the management system, taking account of the graded approach (see paras 5.52–5.58). The management system should include provisions for the information recorded to be checked periodically against the actual state of the radioactive waste, updated as necessary, and then managed to preserve knowledge of the results of waste processing (i.e. pretreatment, treatment and conditioning) and of the status of waste storage and disposal. All important safety related information about radioactive waste management should be retained and controlled.

---

[13] Extended periods of time may apply in cases such as a long-standing industrial operation that generates radioactive waste, the operating and decommissioning periods in the lifetime of a nuclear power plant, the storage of waste awaiting disposal, the disposal of the waste and institutional control during the post-closure period of a disposal facility.

5.66. Records describing the history of the radioactive waste management facility such as data obtained during facility design, construction, commissioning, operation and closure, should also be created and retained. These records should include the following, as appropriate:

(a) Authorizations (e.g. licences, permits, amendments);
(b) Commissioning records;
(c) The safety case and safety assessments;
(d) The environmental impact assessment;
(e) Peer review reports;
(f) Technical specifications and amendments;
(g) Design options, concepts, documents, calculations and drawings;
(h) Records of the facility actually constructed ('as-built' records);
(i) Approved design changes;
(j) Procurement records for structures, systems and components;
(k) Operating procedures;
(l) Records of the implementation, review, updating and maintenance of emergency preparedness and response arrangements, including records of training, exercises, response to actual emergencies, lessons identified and corrective actions implemented;
(m) Waste emplacement plans;
(n) Records generated during facility operation, including records of emplaced waste packages;
(o) Records of assessments, inspections and verifications of processes and activities;
(p) Records of any non-conformances and corrective actions;
(q) Records of the training, experience and qualification of personnel;
(r) Monitoring data;
(s) Records of any incidents, including accidents, that have occurred;
(t) Records of interactions between the operating organization and the regulatory body (e.g. meetings, inspections).

5.67. The licensee should define where and how (e.g. the media to be used) the records are to be stored and this should be documented in the management system. Decisions on record keeping should take account of regulatory requirements and authorization conditions.

5.68. Arrangements should be made to ensure that records are maintained for the appropriate period of time. Retention periods may differ depending on the nature of the waste management facilities and activities, and on the levels of activity and half-lives of the radionuclides involved. The arrangements for maintaining and retaining records should comply with regulatory requirements.

5.69. Information that needs to be retained for an extended period should be subject to regular, periodic and systematic review to examine the implications of any changes that have occurred in the governmental, legal and regulatory framework and in regulatory requirements, and any implications of new organizational, technological and scientific developments.

5.70. Records for a radioactive waste management facility that need to be retained for an extended period should be stored in accordance with regulatory requirements and in a manner that minimizes the likelihood and consequences of loss, damage or deterioration due to events such as fire, flood or other occurrences. The condition of the records should be assessed periodically. When unpredictable events lead to the inadvertent damage to or destruction of records, the condition of any surviving records should be assessed, and the importance and means of their retention and their retention period should be re-evaluated.

5.71. The records should be backed up, retrievable and readable for the specified retention period. Where records are preserved electronically, software and computer hardware should be updated and replaced, and data migrated, as necessary. Irrespective of the storage media used (electronic or other), consideration should be given to the storage of multiple copies in several different locations that have independent systems for the preservation of records.

5.72. If the responsibility for managing radioactive waste is transferred from one organization to another, relevant records and information about the waste and the associated facility should be transferred to the successor organization. The information to be transferred between the organizations should be set out in a document that describes the interfaces and the interactions between the organizations. The information should be provided in a form that the successor organization can read.

5.73. Information on a radioactive waste management facility and its contents may have to be made public, transferred between successive sets of personnel within an organization and made accessible over a long time period. To make it possible for future generations to read, understand and interpret the information, contextual information should be collated, retained and transferred (e.g. information on the governmental, legal and regulatory framework and the regulatory requirements relevant to the facility; the rationale for safety related and optimization decisions; explanations of language and technical terminology;

summaries of scientific understanding; the methods used for collecting, analysing and interpreting measurements) together with the records themselves. The safety case for a facility can be used as a vehicle for the integration and synthesis of this type of information. Consideration should be given to the information, recording media, equipment and systems that will be needed to ensure, as far as possible, that the information will be available in the future. Further information on the preservation of records, knowledge and memory across generations, with particular regard to geological disposal facilities, can be found in Ref. [33].

5.74. The safety case should explain the approach that has been adopted to ensure the safety of the facility. For long term radioactive waste management programmes (and in cases where there is heightened public interest, such as in relation to the siting of disposal facilities), it is especially important to record the reasons why decisions were made and to make this information available and easily traceable for both present and future generations.

MANAGEMENT OF RESOURCES

Provision of resources

5.75. Requirement 9 of GSR Part 2 [5] states that "**Senior management shall determine the competences and resources necessary to carry out the activities of the organization safely and shall provide them.**"

5.76. Resources include individuals (both the number of individuals and their competences), infrastructure, the working environment, knowledge and information, and suppliers, as well as material and financial resources (see footnote 10 of GSR Part 2 [5]).

5.77. The management system should include provisions to ensure that there are sufficient numbers of competent personnel at all levels, that these personnel are suitably qualified and experienced for the tasks allocated to them, and that they understand the safety implications of their work.

5.78. In a typical radioactive waste management process, each step is dependent upon the satisfactory completion of the previous step. Specific training should be provided for personnel involved in the operation of facilities in which radioactive waste is generated or managed, to ensure that they sufficiently understand the processes involved and the interdependencies between all steps in the process of radioactive waste management, and are aware of the potential consequences for safety and the generation of radioactive waste that could arise owing to operator error. Without such understanding, for example, a waste package could be produced that would not meet the acceptance criteria for subsequent storage or disposal.

5.79. Personnel who select, develop and/or implement process technologies for radioactive waste management should be suitably qualified and experienced, and should be trained and competent to perform their assigned tasks. For all stages of radioactive waste management, the operating organization should ensure that operating personnel (including maintenance and technical staff) understand the nature of the waste and its associated hazards, the relevant operating procedures, and the procedures to be followed in the event of incidents, including accidents.

5.80. Human resource plans should be developed and should incorporate measures to ensure the continuous availability of a sufficient number of competent personnel throughout the lifetime of a radioactive waste management facility. For a radioactive waste disposal facility, this includes the period after waste emplacement and the period of active institutional control during the post-closure period.

5.81. Training programmes, procedures and succession plans should be established to ensure that suitable competency is achieved and maintained, and to avoid the potential loss of knowledge, practical experience and technical expertise over time. Senior management is required to ensure that training and refresher training needs are reviewed on a planned basis and updated as required (see para. 4.23 of GSR Part 2 [5]). This includes familiarization with the management system (see para. 4.26 of GSR Part 2 [5]). Further recommendations on the role of the management system in the areas of training, succession planning, and information and knowledge management are provided in GS-G-3.1 [32]. Further information on the preservation of records, knowledge and memory across generations, with particular regard to geological disposal facilities, can be found in Ref. [33].

5.82. Refresher training should be arranged to ensure that personnel adequately understand the implications of changes such as the following:

(a)   Modifications to equipment and materials;
(b)   The installation of new equipment;
(c)   Changes in procedures;
(d)   Changes in technologies for radioactive waste management;
(e)   Any tightening or relaxation of controls (e.g. on the number of waste packages that may be moved at any given time);
(f)   The introduction of additional controls;

(g) Changes in the governmental, legal and regulatory framework;

(h) Changes in regulatory requirements.

5.83. Accumulated experience, including lessons from operations and incidents (including accidents), should be reviewed periodically and used in revising training programmes and in decision making. For long term waste storage facilities and waste disposal facilities, the roles of individuals may change. The knowledge that individuals possess is required to be managed as a resource (see para. 4.27 of GSR Part 2 [5]) and this implies that organizations should take concrete steps to capture the knowledge of their staff.

5.84. Responsibilities, mechanisms and schedules for providing the funds necessary for radioactive waste management should be established in advance, before the funds are needed.

5.85. Senior management should ensure that the management system for radioactive waste management activities includes provisions to deal with funding challenges such as the following:

(a) If the necessary funds cannot be obtained from the waste generator (e.g. as a result of bankruptcy, cessation of business, inadequate financial planning or the transfer of ownership of the waste to other parties), other means of applying the 'polluter pays' principle would need to be considered.

(b) If funds are planned to come from public sources, other demands for such funding may make it difficult to gain access to adequate funds on a timely basis.

(c) It may be difficult to make realistic estimates of costs for radioactive waste management activities that are still in the planning stage and for which little or no experience has been accumulated.

(d) It may be difficult to estimate anticipated costs for activities that will only begin in the long term, because these costs will depend strongly on assumptions made about future inflation rates, bank interest rates and technological developments.

(e) It may be difficult to incorporate appropriate risk and contingency factors into estimates of future costs, owing to the uncertainties associated with future changes in societal demands, political imperatives, public opinion and the nature of unplanned events.

(f) Experience has shown that costs for large projects tend to increase compared with initial estimates (see e.g. Ref. [34]).

(g)   If several organizations are involved in the waste management process, the necessary financial arrangements could be complex and ensuring adequate continuity of funding could be problematic.

5.86. Each operating organization should ensure that adequate commercial arrangements are in place to manage each of the identified waste streams and to ensure that these arrangements are likely to endure for the period necessary to complete the waste management programme. The government and the regulatory body should ensure that adequate contingency planning is included in these arrangements. If the financial arrangements prove to be inadequate, then the government may have to take measures to ensure that the waste continues to be managed safely.

MANAGEMENT OF PROCESSES AND ACTIVITIES

Management of processes and activities

5.87. Requirement 10 of GSR Part 2 [5] states that "**Processes and activities shall be developed and shall be effectively managed to achieve the organization's goals without compromising safety.**"

5.88. GSR Part 2 [5] states:

"4.28. Each process shall be developed and shall be managed to ensure that requirements are met without compromising safety. Processes shall be documented and the necessary supporting documentation shall be maintained. It shall be ensured that process documentation is consistent with any existing documents of the organization. Records to demonstrate that the results of the respective process have been achieved shall be specified in the process documentation.

"4.29. The sequencing of a process and the interactions between processes shall be specified so that safety is not compromised. Effective interaction between interfacing processes shall be ensured. Particular consideration shall be given to interactions between processes within the organization, and to interactions between processes conducted by the organization and processes conducted by external service providers.

"4.30. New processes or modifications to existing processes shall be designed, verified, approved and applied so that safety is not compromised.

Processes, including any subsequent modifications to them, shall be aligned with the goals, strategies, plans and objectives of the organization.

"4.31. Any activities for inspection, testing, and verification and validation, their acceptance criteria and the responsibilities for carrying out such activities shall be specified. It shall be specified when and at what stages independent inspection, testing, and verification and validation are required to be conducted.

"4.32. Each process or activity that could have implications for safety shall be carried out under controlled conditions, by means of following readily understood, approved and current procedures, instructions and drawings. These procedures, instructions and drawings shall be validated before their first use and shall be periodically reviewed to ensure their adequacy and effectiveness. Individuals carrying out such activities shall be involved in the validation and the periodic review of such procedures, instructions and drawings."

5.89. Many processes support waste minimization, handling, pretreatment, treatment, conditioning, transport, storage and disposal. The management system should provide assurance that these processes comply with all applicable requirements, and should encourage the application of the principle of carrying out work correctly the first time.

5.90. All radioactive waste management processes should be specified, and individual 'process owners' should be appointed by senior management. The authorities and responsibilities of process owners should be documented. The following processes should be considered, as appropriate:

(a) Research and development;
(b) Commissioning, calibration and testing of new equipment;
(c) Commissioning of new processes and activities;
(d) Safety case development, including safety and performance assessments;
(e) Facility design and optimization, and facility construction;
(f) Environmental protection, environmental monitoring and surveillance;
(g) Control of products (e.g. waste containers, waste packages);
(h) Providing for traceability of waste;
(i) Condition monitoring, particularly during any long term storage and when radioactive waste is in a disposal facility prior to the closure of that facility;
(j) Retrieval of waste from waste management facilities;
(k) Knowledge and information management, including the control, transfer and keeping of records;

(l) Development of waste acceptance criteria, acceptance of waste and transfer of responsibility;
(m) Radiation protection;
(n) Ensuring legal compliance;
(o) Risk management;
(p) Applying a graded approach to the application of management system requirements;
(q) Process management;
(r) Decision making;
(s) Communication with interested parties;
(t) Human resources management;
(u) Procurement;
(v) Management of organizational change and resolution of conflicts;
(w) Documentation of the management system and measurement, assessment and improvement of the management system;
(x) Managing interactions between the management system processes;
(y) Emergency preparedness and response;
(z) Post-closure institutional control and monitoring of a disposal facility.

5.91. The possibility of human error should be taken into account when defining processes and activities. The processes and activities should be planned so as to minimize potential human errors.

Development of processes

5.92. The management system should include procedures for the design of radioactive waste management processes. The design of these processes should take into account the hierarchy of hazard controls; that is (in order of decreasing effectiveness), hazard elimination, hazard substitution, engineered controls, administrative controls and the use of personal protective equipment. Examples of hazard elimination include minimizing the generation of waste and reusing or recycling a disused sealed radioactive source. Examples of hazard substitution include the storage of vitrified high level waste instead of liquid high level waste and the storage of uranium in oxide form instead of as uranium hexafluoride. Examples of engineering controls include the use of shielding or remote handling technologies. Examples of administrative controls include training, supervision and operating procedures. Personal protective equipment includes protective clothing and face masks to avoid skin contamination and internal contamination.

5.93. The design of processes for the predisposal management of radioactive waste should take into account the detailed sequence of tasks involved, and issues relating to the specific work processes and products (e.g. waste packages), including, for example, the following:

(a) The planning and implementation of a radiation protection programme[14] for facility operation, including the use of protective clothing and shielded equipment and facilities;

(b) The use of special handling equipment, tools and techniques for the emplacement of waste packages in, and their retrieval from, storage facilities;

(c) Testing and assay requirements (e.g. equipment, methods, materials);

(d) The design of non-destructive systems and chemical analysis methods for waste characterization;

(e) The need for waste pretreatment, treatment and conditioning;

(f) The design of waste containers and waste packages on the basis of detailed specifications for materials, geometry, mechanical properties, sealing and containment capability, stability, robustness and durability (see IAEA Safety Standards Series Nos SSG-40, Predisposal Management of Radioactive Waste from Nuclear Power Plants and Research Reactors [36], and SSG-41, Predisposal Management of Radioactive Waste from Nuclear Fuel Cycle Facilities [37]);

(g) The design of waste transport containers and waste storage facilities for use before development of a disposal facility, taking into account the uncertainty in the possible design of the disposal facility;

(h) The length of time that waste is to be stored, taking into account the characteristics of the waste, the waste package and the store;

(i) Events and processes that could lead to the degradation and possible failure of waste packages in storage;

(j) The design of methods for examining waste packages that may have degraded while in storage;

(k) The need to inspect, move, repair and/or re-engineer waste packages in storage;

(l) The need for equipment to be maintained and replaced during operations and the possible need for any specialized equipment in the future;

(m) The need to maintain storage facility buildings, potentially over a very long lifetime.

---

[14] Requirements for the protection of workers, including the establishment of a radiation protection programme are established in IAEA Safety Standards Series No. GSR Part 3, Radiation Protection and Safety of Radiation Sources: International Basic Safety Standards [35].

5.94. The design of processes for the disposal of radioactive waste should take into account the detailed sequence of tasks that will be involved and issues relating to the specific work processes, including, for example, the following:

(a) Planning and implementation of site investigations while minimizing disruption to the integrity of the natural environment and the host geological formation;
(b) Planning, use and sealing of site investigation boreholes that if not sealed might affect the safety of the disposal system;
(c) Design and construction of facilities (e.g. near surface disposal vaults, the excavation of underground cavities) while minimizing damage to the natural environment and the host geological formation;
(d) Planning and implementation of a radiation protection programme for facility operation, including the use of protective clothing and shielded equipment and facilities;
(e) The use of special handling equipment, tools and techniques for the emplacement of waste in disposal facilities;
(f) The installation of engineered barriers (e.g. buffers, backfills, seals, closure components);
(g) Protection of waste packages from degradation (e.g. due to rock fall or corrosion) before the facility is closed;
(h) Monitoring activities.

5.95. Experiments, pilot scale tests and commissioning procedures carried out to support the design of processes that are to be implemented full scale in radioactive waste management should have the following aims:

(a) To provide assurance that it will be possible to quantify, either by direct measurement or by process control, the important waste parameters and characteristics (e.g. mass of fissile material, isotopic composition, chemical composition and physical state, decay heat) necessary to control the processes;
(b) To determine the process variables that are important to the acceptability of the end product and the parameters that are potentially important to safety.

5.96. Investigations that are performed to support the design of processes and that employ simulated waste or simulated waste constituents should be focused on ensuring the following:

(a) That the compositions of simulated waste are, as far as possible, representative of the actual waste to be processed;

(b)   That any anticipated conditions that might result in a significant reduction in the quality of waste packages are considered.

5.97. Any previously unrecognized variations (e.g. in the composition of waste streams) should be considered to determine whether they necessitate adjustment of the design of processes or of the specifications for the materials currently being used for waste conditioning.

5.98. For long term waste management activities, future infrastructural needs should be specified and the operating organization should develop plans to ensure that these needs will be met. In such planning, consideration should be given to the continuing need for the following:

(a)   Support services;
(b)   Spare parts for equipment that might cease to be manufactured during the long operational period of the facility;
(c)   Equipment upgrades to meet new regulatory requirements or to make operational improvements;
(d)   Provisions to address the evolution and obsolescence of computer hardware and software.

5.99. Procedures should be established to ensure that the status of waste being processed and the status of equipment, tools, materials and other items important to safety are known and controlled at all times in order to ensure the following:

(a)   Required tasks, inspections or tests are not inadvertently omitted;
(b)   Non-conforming equipment is not installed, used or relied on;
(c)   Tools and items of test equipment that are possibly damaged, defective or out of calibration are not used;
(d)   Non-conforming materials and items (e.g. waste, immobilizing agents, waste containers) are identified and segregated and are not processed further until the non-conformance is resolved.

5.100.   The development of management processes for radioactive waste management activities should take into account the following aims:

(a)   To ensure the continuity of control of the waste and waste management activities;
(b)   To maintain links and relationships between organizations, where more than one organization is involved;

(c) To plan for the long duration of the waste management activities, where appropriate;

(d) To ensure that safety will be maintained throughout the long lifetime of a disposal facility;

(e) To ensure the retention of knowledge of the waste and of the waste management activities.

5.101. The management processes should be suitable for the relevant stage of the waste management programme. The development of each process should ensure that the requirements and the risks relating to the waste management activities have been identified, and that interfaces and interactions with other processes have been taken into consideration.

Special processes

5.102. Special processes are processes for which one or both of the following apply:

(a) The output from the process depends strongly on the control of the process, the skill of the operating personnel or both;

(b) It is not possible to fully confirm the conformity of the output with the specified acceptance criteria by inspection or testing after the process has been conducted (e.g. the welding of lids onto certain types of waste containers, the backfilling of a radioactive waste disposal facility).

5.103. Special processes in the predisposal management of radioactive waste include the following:

(a) The use of certain analytical methods and sampling protocols in waste characterization and process control;

(b) Non-destructive examination and testing of structures, systems and components (e.g. waste containers);

(c) Welding;

(d) Heat treatment;

(e) Painting and coating of containers of waste that generate high radiation levels;

(f) Non-destructive examination and testing of waste (e.g. radiography in real time, gamma and neutron measurement techniques);

(g) Corrective actions for waste containers and waste packages that do not comply with specified requirements (e.g. closure welding of lids on overpacks).

5.104. Special processes in the disposal of radioactive waste include the following:

(a) Processes involving remote handling methods (e.g. for controlled emplacement of waste packages and backfill materials in the presence of high radiation levels);
(b) Certain waste emplacement activities (e.g. emplacement of large spent fuel containers);
(c) The construction, installation, maintenance and monitoring of structures, systems and components (e.g. engineered barriers);
(d) The retrieval, repair and relocation of waste packages if problems arise after they have been emplaced.

5.105. Special processes should be witnessed by suitably qualified and experienced personnel and shown to be effective for the conditions in which they will be applied, and any limitations should be documented.

5.106. The validation of non-destructive gamma or neutron techniques for the analysis of radioactive waste should involve the following:

(a) Use of empirical data to validate algorithms for measuring the activity of radionuclides;
(b) In developing the method or in calibrating the equipment, use of reference materials that have the same attenuation properties and moderating properties as the radioactive items to be measured (e.g. waste or waste packages);
(c) Quantification of the uncertainties associated with the analysis of the waste.

5.107. Special processes should be performed by suitably qualified and experienced personnel and conducted in accordance with approved procedures. Appropriate records of special processes should be made and retained. Where industry standards apply to special processes, the requirements of such standards should be complied with. When any changes are made (e.g. to samples and conditions, methods, equipment, or qualification of personnel), the special processes should be revalidated.

Inspection and testing of processes

5.108. Inspection and testing are important for controlling work processes and should be planned, documented, executed and recorded to ensure that important parameters of waste management processes are controlled, that waste and waste packages meet design specifications and acceptance criteria, and that disposal facility conditions at the time of waste emplacement meet the design specifications and expected initial state. Inspection criteria should be specified for each inspection and testing step in radioactive waste management.

5.109.   Inspections carried out as part of radioactive waste management should include the following, as appropriate:

(a)   Inspection at source of items important to safety for which the quality is difficult to verify upon receipt;
(b)   Inspection and testing, as appropriate, on receipt of items important to safety (e.g. waste packages), including verification of related certification and documentation;
(c)   Inspection of installed items that are important to safety, including witnessing of equipment or system operational tests;
(d)   Appropriate acceptance inspections to validate structures, systems and components;
(e)   Inspection of radioactive waste treatment processes;
(f)   Inspection of processes (e.g. by non-destructive analysis or real time radiography) used for determining whether waste forms are suitable and can be accepted;
(g)   Inspection of radioactive waste processing facilities and activities;
(h)   Final inspection of waste forms, waste containers and waste packages destined for storage;
(i)   Inspection of radioactive waste containers and waste packages that are designed to comply with the requirements established in SSR-6 (Rev. 1) [8];
(j)   Periodic, non-invasive inspection of the integrity and identification of radioactive waste in storage;
(k)   Inspection of the radioactive waste management facility construction processes;
(l)   Inspection of the facility before radioactive waste is accepted for receipt;
(m)   Inspection of radioactive waste emplacement and engineered barrier installation processes;
(n)   Inspection (e.g. by non-destructive analysis or real time radiography) of radioactive waste destined for disposal (e.g. on receipt at the radioactive waste disposal facility, during storage awaiting disposal), including either comprehensive inspection or random sampling inspection;
(o)   Regular inspection to verify the operability of equipment or systems used for the prevention, detection or mitigation of accidents.

5.110.   For inspections and tests designed to verify the characteristics of a radioactive waste container or waste package (e.g. geometry, mechanical properties, sealing and containment capability, stability, robustness, durability; see SSG-40 [36] and SSG-41 [37]), methods should be used that have been demonstrated to be effective on the materials to be tested and test conditions should be representative of (or more severe than) the conditions that will be encountered during storage or disposal.

Validation and verification of processes

5.111. Work processes should be validated to ensure that they are appropriate for achieving their intended function. The results and outputs of the processes implemented should be verified to determine whether they are of the necessary quality.

5.112. The validation of work processes should be performed in accordance with documented and approved procedures, and the results should be reported. Validation of work processes, where feasible, should include the following:

(a) Identification of the process variables that should be controlled to ensure the adequacy of radioactive waste management activities;
(b) Establishment of the limits or tolerances for the process variables;
(c) Identification of adequate control methods for the process variables, including the frequency of sampling and testing of waste forms and packages.

5.113. The verification of process outputs should include establishing and implementing an appropriate testing and monitoring programme with which to verify the quality of the outputs at each stage of the waste management programme (e.g. radioactive substances discharged, materials cleared from regulatory control, waste, waste forms and waste packages for storage and disposal). The rationale for the design of the testing and monitoring programme should be documented.

5.114. Radioactive waste management processes should include appropriate 'hold points' at which the acceptability of important results or outputs should be verified. Procedures should specify that the process should not proceed beyond hold points until designated personnel have given their approval. Independent verification should be provided, where appropriate, commensurate with the safety significance of the activity (i.e. in accordance with a graded approach). Where appropriate, hold points may be waived if a satisfactory justification on grounds of safety or quality is documented and approved.

5.115. If reports and records relating to the manufacture of waste containers and the conditioning of waste to produce waste packages do not demonstrate that the waste packages meet the acceptance criteria for disposal (e.g. because the waste packages were produced before waste acceptance criteria for a disposal facility were set), then further assessment should be undertaken to determine whether the waste packages meet the waste acceptance criteria for disposal. If the waste packages do not meet the waste acceptance criteria for disposal, the need to rework the packages should be considered.

5.116.   Appropriate reports and records of the validation of work processes and the verification of process outputs should be kept. Reports and records of the verification of process outputs should be made available to all subsequent waste processors, storage facilities, consignors, radioactive waste disposal facilities and the regulatory body.

Optimization of processes

5.117.   The management system should include a process and procedures for optimizing work processes, and these should be applied before and throughout each stage of facility development in an iterative, systematic and transparent manner. Optimizing waste management work processes should include the following activities, as appropriate:

(a)   For the production of waste packages: providing guidance and training to waste generators on the waste acceptance criteria at the relevant radioactive waste disposal facility as early as possible.
(b)   For site characterization: maximizing understanding of baseline conditions and knowledge drawn from non-invasive investigations of a site, in addition to the use of selective and justified invasive methods such as borehole investigations.
(c)   For environmental impact assessment: minimizing disturbance of the environment.
(d)   For facility design: coordinating the interaction between the activities and teams involved in facility design, site characterization and safety case development.
(e)   For construction: selecting techniques and equipment that will minimize disturbance to the site, including the host geological formation, especially close to major discontinuities and zones of structural weakness.
(f)   For operation of storage facilities and disposal facilities: emplacing waste packages intact and without significant damage and in accordance with any waste emplacement strategy included in the safety case for the facility.
(g)   For operation and closure of a disposal facility: emplacing backfill and installing seals at the correct densities to achieve the intended hydraulic conductivities at the necessary locations in accordance with the design

requirements; installing the cap over a near surface disposal facility so as to minimize the flow of water to the waste.

(h) For the post-closure period: arranging for archiving of relevant materials and information.

Management of non-conformances

5.118.   The management system should include a process and procedures to deal with non-conforming items. The process and procedures should include the following actions:

(a) Timely identification, reporting and documentation of counterfeit or fraudulent items and other non-conformities;

(b) Segregation of non-conforming items to prevent them from being used or transferred to another organization before the non-conformance is resolved;

(c) Clear identification of non-conforming items and process equipment (e.g. tagging, labelling, stickers, marking);

(d) Assessment of the condition of non-conforming items, resolution of non-conformances (e.g. rework, repair, reject), and determination of the causes for non-conformances so that corrective actions can be taken to prevent the non-conformances from recurring;

(e) Appropriate follow up, as necessary, to evaluate the effectiveness of corrective actions.

5.119.   The consequences of the non-conformance of an item should be evaluated to assess whether the item can be accepted, or whether it should be reworked or repaired to bring it back into conformity. If none of these options is practicable, or if the item is found to be a counterfeit or fraudulent item, the item should be rejected. The management system should include a process and procedures that describe how such rejected items are to be managed. In the case of a waste package for which neither repair nor rejection is a viable option, consideration should be given to reworking the package, overpacking or taking other measures (e.g. using a new waste container) to bring the package into compliance with the waste acceptance criteria. Any non-conformance that is important to safety that is found after the emplacement of the waste (e.g. a design fault, defective package material or damage affecting the integrity of the package) should be rectified as far as possible. If the non-conformance cannot be rectified, its impact on safety should be subjected to a detailed analysis, which should be used as appropriate to optimize the situation. For example, if a waste package that has been emplaced in a facility is found to have been damaged, its effect on safety should be assessed and the results of the assessment should be used, together with other information on potential management options, to decide on the management option to be implemented (e.g. it may not be possible to re-weld a damaged waste package, but it might be possible and justified to provide an overpack).

5.120. Non-conformance data should be analysed periodically to identify trends (e.g. in the quality of waste containers and waste packages), and these analyses should be reported to the responsible manager for review. Corrective actions should be initiated to remove or eliminate the underlying causes of the non-conformances where these are important to safety.

Management of the supply chain

5.121. Requirement 11 of GSR Part 2 [5] states that "**The organization shall put in place arrangements with vendors, contractors and suppliers for specifying, monitoring and managing the supply to it of items, products and services that may influence safety.**"

5.122. GSR Part 2 [5] also states:

"4.33. The organization shall retain responsibility for safety when contracting out any processes and when receiving any item, product or service in the supply chain[11].

"4.34. The organization shall have a clear understanding and knowledge of the product or service being supplied[12]. The organization shall itself retain the competence to specify the scope and standard of a required product or service, and subsequently to assess whether the product or service supplied meets the applicable safety requirements.

"4.35. The management system shall include arrangements for qualification, selection, evaluation, procurement, and oversight of the supply chain.

"4.36. The organization shall make arrangements for ensuring that suppliers of items, products and services important to safety adhere to safety requirements and meet the organization's expectations of safe conduct in their delivery.

"[11] The supply chain, described as 'suppliers', typically includes: designers, vendors, manufacturers and constructors, employers, contractors, subcontractors, and consigners and carriers who supply safety related items. The supply chain can also include other parts of the organization and parent organizations.

"[12] The capability of the organization to have a clear understanding and knowledge of the product or service to be supplied is sometimes termed an 'informed customer' capability."

5.123. In cases where there are long time periods involved in radioactive waste management, the responsible organization should plan how it will manage the availability and quality of equipment and the procurement of any structures, systems or components that need to be replaced. The organization should monitor suppliers so that they do not cease operation without prior warning, and ensure that there is a diversity of supply. The organization should ensure that it has sufficient spare parts. In some instances, research and development may be necessary to provide an advance warning of the potential failure of equipment or structures, systems or components, or to identify potential replacements. Financial arrangements should be put in place to accommodate long term needs, and procurement plans should be consistent with these.

5.124. Organizations that commission services, items or processes should have sufficient capabilities in house to act as an 'intelligent customer' (e.g. see IAEA Safety Standards Series Nos GS-G-3.5, The Management System for Nuclear Installations [38] and GSG-12, Organization, Management and Staffing of the Regulatory Body for Safety [39]). Services, items and processes supplied by other organizations should be controlled through contractual arrangements that, for example, include the following:

(a) Management system requirements;
(b) Specifications for services, items and processes, as appropriate;
(c) Validation and verification criteria;
(d) Regulatory requirements;
(e) Resource requirements and constraints.

5.125. Potential suppliers should be provided with a clear description of the items and services to be supplied. The process to be used for evaluating proposals from potential suppliers and for selecting suppliers should also be made available. Use of a list of approved and preferred suppliers prevents redundant effort in procurement and helps ensure consistency of acceptance of the suppliers. Acceptability of proposals and suppliers should be based on appropriate selection criteria, such as capability to meet purchasing requirements, the qualifications and experience of the individuals identified to manage and conduct work, the proposed approach for supplying the items and services, the track record of the organizations being subcontracted (especially in terms of safety performance), client and third party audits of suppliers and subcontractors, costs and the acceptability of any gaps in the supplier's proposals. The details of the procurement process — including the reasons for selecting the chosen supplier and the contract documentation — should be recorded.

5.126.   In planning for procurement, consideration should be given to the availability and quality of equipment (e.g. monitoring instrumentation), materials and other items important to safety over the extended periods of waste storage and disposal. Consideration should also be given to the financial arrangements and controls that may be necessary over such extended periods of time.

5.127.   The management systems of organizations in the supply chain should be reviewed and revised as necessary so that they can be accepted before work commences. The management systems of organizations in the supply chain should include arrangements for oversight of contractors by the organization that procures the work. Oversight of contractors should include surveillance, inspection of activities, ongoing monitoring, periodic review by experts, acceptance of plans and deliverables, and review of changes to activities.

Application of the management system to the development of the safety case for radioactive waste management facilities and activities

5.128.   Requirement 13 of GSR Part 5 [3] states:

> **"The operator shall prepare a safety case and a supporting safety assessment. In the case of a step by step development, or in the event of modification of the facility or activity, the safety case and its supporting safety assessment shall be reviewed and updated as necessary."**

5.129.   Requirement 12 of SSR-5 [4] states:

> **"A safety case and supporting safety assessment shall be prepared and updated by the operator, as necessary, at each step in the development of a disposal facility, in operation and after closure. The safety case and supporting safety assessment shall be submitted to the regulatory body for approval. The safety case and supporting safety assessment shall be sufficiently detailed and comprehensive to provide the necessary technical input for informing the regulatory body and for informing the decisions necessary at each step."**

5.130.   The senior management of the operating organization of a radioactive waste management facility is responsible for developing, reviewing and maintaining a safety case to provide the basis for decisions on facility design, construction, commissioning, operation and decommissioning or closure, as appropriate. Recommendations on the development of safety cases and supporting safety assessments for facilities for predisposal management of radioactive waste and for radioactive waste disposal facilities are provided in GSG-3 [16] and SSG-23 [17], respectively.

5.131. The management system relates to the safety case in several ways as follows:

(a) The management system should ensure that the safety case has been appropriately prepared, reviewed and updated by suitably qualified and experienced personnel.

(b) The management system should ensure that all relevant requirements relating to the safety case and the safety of the facility are met. Furthermore, the management system should provide confidence that the relevant requirements will continue to be met throughout all steps in the lifetime of the waste management facility.

(c) The management system should include processes and procedures to ensure the quality of all activities associated with the safety case, such as data collection and safety assessment modelling. The safety case for a disposal facility may need to deal with particular uncertainties relating to the length of the assessment period (e.g. thousands of years) and the need to consider and model the behaviour of the natural system at the site and its evolution; where relevant, the management system should include specific processes and procedures for the management of such uncertainties.

(d) The management system should identify the process for developing and applying waste package specifications and waste acceptance criteria consistent with, and derived from, the relevant safety cases (including, as appropriate, the safety case for the subsequent waste management facility), in accordance with Requirement 12 of GSR Part 5 [3] and Requirement 20 of SSR-5 [4].

(e) The management system should include processes and procedures for the periodic review and updating of the safety case (e.g. to take account of operational experience, new or revised standards and other newly available information).

(f) The safety case should acknowledge the existence of unresolved issues and provide guidance for work to resolve these issues in future development stages. The management system should include processes and procedures for initiating and managing tasks aimed at addressing unresolved issues. The safety case should also enable the parties involved to judge the level of safety provided by the waste management facility throughout its development and as new information is obtained.

(g)   The management system should include processes and procedures to ensure that personnel involved in the development of the safety case are suitably qualified and experienced and that they have an understanding of the waste management system and its associated risks and of the process used for the review of the safety case.

5.132.   The safety case is part of the information to be provided to the regulatory body for review and approval in support of facility authorization. The management system should include processes and procedures for the provision of appropriately detailed and comprehensive information to the regulatory body to fulfil regulatory requirements throughout the lifetime of the facility.

5.133.   The management system should include arrangements for establishing, as appropriate, waste package specifications, waste acceptance criteria and other conditions to be applied at the facility. The management system should also include arrangements (e.g. by establishing technical specifications based on the safety case and safety assessments) to ensure that the facility is designed, constructed, operated and decommissioned or closed, as appropriate, in accordance with the safety case and the authorization.

5.134.   The management system should include processes and procedures for the retention and archiving of information relevant to the safety case and the safety assessments (see Requirement 15 of GSR Part 5 [3] and Requirement 14 of SSR-5 [4]), as well as records of facility operation and inspections that demonstrate compliance with regulatory requirements, technical specifications, waste package specifications, waste acceptance criteria and other conditions. Such information and records should be retained with other important records, as described in paras 5.64–5.74.

5.135.   The following aspects should be taken into account in the management system for the development of a safety case:

(a)   The need for internal and external audits, as appropriate, of information and activities relating to the safety case, to determine the adequacy of the management system and its implementation;
(b)   The need to demonstrate the suitability of personnel involved in the preparation of the safety case (e.g. conducting safety assessments, undertaking safety case reviews) by documenting their qualifications and

experience, and by providing further relevant training (e.g. through their participation in international projects);

(c)   The need to take into account the views of interested parties on the safety case;

(d)   The need to ensure consideration of applicable international safety standards in the development of the safety case;

(e)   The need for robust processes and procedures to ensure the quality of data, models, software and calculations used to underpin the safety case;

(f)   The need to develop and maintain competence with regard to the safety case, both within the operating organization and within the regulatory body, over the whole project time frame.

5.136.   The management system should, in particular, include processes and procedures for ensuring the traceability and transparency of the safety case, for research and development, for the treatment of uncertainty and for the integration of safety case development with the design and optimization of the facility.

Traceability and transparency

5.137.   A coherent referencing system supporting the safety case should be established. This should include structured information on when, on what basis and by whom decisions and assumptions were made, on how these decisions and assumptions were implemented, on what modelling tools were used, and on the original sources of the information used in the safety case.

5.138.   The management system should include a transparency policy that provides for and ensures openness, communication and accountability. Because of the longevity of the hazards associated with some radioactive waste management facilities (particularly disposal facilities for long lived radioactive waste), the need for transparency when interacting with interested parties is particularly important.

5.139.   The safety case and safety assessment should be documented in a clear, open and unbiased way that describes the safety features of the radioactive waste management system and any associated uncertainties. The aim should be to provide a clear picture of what has been done in the safety assessments, what the results and uncertainties are, why the results are as they are and what the key issues are, in order to inform decision makers. To increase transparency, the documentation of the safety case should, to the extent possible, taking account of circumstances relating to security or commercial confidentiality, be made available to the public and should be prepared taking account of the graded approach and in a manner and at a level of detail that is suitable for the intended audience.

Research and development

5.140. In relation to the predisposal management of radioactive waste, para. 3.10 of GSR Part 5 [3] states that "the regulatory body, where appropriate, may undertake research". Paragraph 5.12 of GSR Part 5 [3] states that "the safety assessment has to be reviewed and updated...[w]hen there are significant developments in knowledge and understanding (such as developments arising from research...)".

5.141. In relation to the disposal of radioactive waste, para. 2.32 of GSR Part 1 (Rev. 1) [18] states that "The government shall make provision for appropriate research and development programmes in relation to the disposal of radioactive waste, in particular programmes for verifying safety in the long term."

5.142. In relation to the disposal of radioactive waste, para. 3.13 of SSR-5 [4] states:

> "The operator has to conduct or commission the research and development work necessary to ensure that the planned technical operations can be practically and safely accomplished, and to demonstrate this. The operator likewise has to conduct or commission the research work necessary to investigate, to understand and to support the understanding of the processes on which the safety of the disposal facility depends."

5.143. To address these requirements, the management systems of the relevant organizations should include provisions for the development, review and maintenance of a high level document that describes the research and development programme for the establishment of a radioactive waste disposal facility. The identity of the relevant organizations will depend on national arrangements, but it is typically the operating organization that takes the leading role in conducting or commissioning the research and development. The research and development programme document should describe conducted, ongoing and planned research relevant to the safety of the facility and should integrate the research outputs that could be used to support the safety case; an example is provided in Ref. [40]. The research and development programme should address the scheduling of activities and how the research and development programme is connected to the future development of the safety case and safety assessments, and to the facility design and waste management activities.

5.144. Because of the longevity of some radioactive waste management facilities, especially disposal facilities, research may need to be initiated well in advance of operation and be capable of assessing long term behaviour and, consequently, should inform the development of waste acceptance criteria. The operating organization should take account of experience and lessons from other States and should conduct or commission the research work necessary to investigate and understand the waste management system, the structures, systems and components on which the safety of the facility depends, and the events and processes that could affect its performance. This should include obtaining all the data necessary for safety assessment, including assessing the suitability of the materials used in the facility.

5.145.   The research and development activities involved in developing and assessing the safety of a radioactive waste disposal facility can be very wide ranging and may be conducted both in the laboratory and in the field. The provisions in the management system that manage research and development activities should be capable of dealing with a wide range of studies and conditions (e.g. surface and underground laboratory conditions, investigations in natural environment) and timescales (e.g. from days to years or decades). The management system should recognize that there will always be uncertainty in the results from research and development activities and should include processes and procedures for dealing with these uncertainties.

Treatment of uncertainties

5.146.   The management system should ensure that uncertainties (e.g. in assessing the behaviour of natural systems and engineered systems) are, as far as possible, identified and that the basis for their quantification, where this is considered appropriate, is clearly documented. Recommendations on the management of uncertainties in the preparation of the safety case and safety assessment for predisposal management of radioactive waste are provided in GSG-3 [16] and, for the disposal of radioactive waste, in SSG-23 [17]. Further information is provided in Refs [41, 42].

5.147.   At any particular stage in assessing the safety of a radioactive waste management facility, there might not be sufficient information available to provide the necessary level of confidence. This might be the case if the information used in the safety assessment is derived from the following:

(a)   Generic (i.e. not site specific) studies;
(b)   Estimated values;
(c)   Extrapolated values;
(d)   Studies that were conducted for other purposes.

5.148. Uncertainties associated with insufficient information can be addressed by using appropriate approaches to safety assessment, including conservative deterministic safety assessment calculations, sensitivity and uncertainty analyses, and probabilistic risk assessments. Even if these approaches are used, further research and uncertainty management may still be necessary.

5.149. The compilation and use of data should be clearly described, justified and recorded. As more data are collected — for example, during a site characterization programme — the level of reliance on generic studies and on estimated and extrapolated values should decrease, and with appropriate management of uncertainties, the level of confidence in the safety case should increase.

5.150. When statistical data that have been compiled on a large scale (e.g. regional data on geological or hydrogeological characteristics) are used, special consideration should be given to how such data can be applied to the particular site of the disposal facility and its immediate surroundings. Similarly, special consideration should be given to how data collected on a small scale (e.g. in the laboratory) can be applied to the full scale operation of the disposal facility. The management system should address any issues associated with the scaling of data.

5.151. When computer software and models are used to support radioactive waste management activities, the management system should ensure that appropriate model and software verification and validation are carried out, taking into account the uncertainties associated with modelling the long term behaviour of disposal systems.

Optimization of radioactive waste management

5.152. The overall process of radioactive waste management should be optimized. In addition, each stage of radioactive waste management, including predisposal management activities and their design, and throughout the lifetime of predisposal management facilities and disposal facilities (i.e. during site selection and characterization, and facility design, construction, operation and decommissioning or closure, as appropriate) should be optimized. The management system should include a process and procedures for considering a wide range of technical, socioeconomic and environmental factors, including using the safety case and safety assessment, to guide stepwise and iterative decision making on the selection of options in each circumstance.

Application of the management system to all steps in the management of radioactive waste

5.153.   Requirement 7 of GSR Part 5 [3] states that "**Management systems shall be applied for all steps and elements of the predisposal management of radioactive waste.**" Furthermore, para. 3.24 of GSR Part 5 [3] states:

"To ensure the safety of predisposal radioactive waste management facilities and the fulfilment of waste acceptance criteria, management systems are to be applied to the siting, design, construction, operation, maintenance, shutdown and decommissioning of such facilities and to all aspects of processing, handling and storage of waste."

5.154.   Requirement 25 of SSR-5 [4] states (footnote omitted):

"**Management systems to provide for the assurance of quality shall be applied to all safety related activities, systems and components throughout all the steps of the development and operation of a disposal facility. The level of assurance for each element shall be commensurate with its importance to safety.**"

Generation and management of radioactive waste away from facilities primarily intended for radioactive waste management

5.155.   Some predisposal management activities take place away from facilities that are primarily intended for radioactive waste management. For example, some radioactive waste is managed where it is generated (e.g. in hospitals or in industrial, agricultural or research facilities). Some radioactive waste management activities are conducted using mobile equipment and facilities. GSR Part 5 [3] and IAEA Safety Standards Series No. SSG-45, Predisposal Management of Radioactive Waste from the Use of Radioactive Material in Medicine, Industry, Agriculture, Research and Education [43] address various approaches to managing radioactive waste, including 'delay and decay', 'concentrate and contain', 'isolate' and 'dilute and disperse', and address the need for minimization and characterization of radioactive waste. SSG-45 [43] provides specific guidance on the application of the management system to activities in medicine, industry, agriculture, research and education that generate radioactive waste.

Siting of radioactive waste management facilities

5.156. Siting is an important process for radioactive waste management facilities. Site characterization supports siting decisions, but also continues after site selection. Site characterization is an important process that contributes to the development of sufficient understanding of the site and that supports development of the safety case for the facility. This is especially the case for disposal facilities because the site forms part of the disposal system and contributes to the safety of disposal. The siting process for radioactive waste management facilities should be clearly defined, transparent and agreed among interested parties, as appropriate. The process should enable the necessary site characterization activities and safety case development work to inform decisions regarding the selection of a site and the design and development of the facility. Siting decisions should be evidence based, and consideration should be given to the views of interested parties. Recommendations on the siting of different types of radioactive waste management facility are provided in the following Safety Guides:

(a) IAEA Safety Standards Series No. WS-G-6.1, Storage of Radioactive Waste [44];
(b) IAEA Safety Standards Series No. SSG-1, Borehole Disposal Facilities for Radioactive Waste [45];
(c) IAEA Safety Standards Series No. SSG-14, Geological Disposal Facilities for Radioactive Waste [46];
(d) IAEA Safety Standards Series No. SSG-29, Near Surface Disposal Facilities for Radioactive Waste [47].

5.157. The management system for radioactive waste management should include a process and procedures for developing and implementing a reasoned, scientifically based site characterization programme. The site characterization programme should be designed to collect information, as necessary, to assess and demonstrate safety and inform the design of the facility. The management system should include a process and procedures for the periodic review and modification of the site characterization programme as data are collected.

5.158. In accordance with the graded approach, the scale and duration of the site characterization programme should reflect the magnitude of the hazard posed by the waste to be managed and the complexity of the situation. For example, the site characterization programmes for a small waste store and for a borehole disposal facility for a small inventory of disused sealed radioactive sources (see SSG-1 [45]) might be considerably less extensive than that for a geological disposal facility for high level radioactive waste, especially one for a wide range of waste streams (see SSG-14 [46]).

5.159.   The management system should include a process and procedures for the operating organization, particularly for that of a disposal facility, to ensure that site characterization does not unduly disturb the surrounding environment (e.g. the hydrogeochemical environment). This should include monitoring procedures, as appropriate, to determine the extent of disturbance by site characterization activities.

5.160.   A systematic process should be defined and applied for collecting and analysing site characterization and environmental data in support of site selection, the development of the safety case, facility design and, where needed, the development of an environmental impact assessment. Such data should be collected prior to facility construction, during construction, during operation and after the closure of a disposal facility as required by the safety case and any applicable regulations. Recommendations on the preparation of an environmental impact assessment are provided in IAEA Safety Standards Series No. GSG-10, Prospective Radiological Environmental Impact Assessment for Facilities and Activities [48].

5.161.   All new data should be collected in accordance with the management system. Written procedures should be developed and used to ensure that data collected are of high quality, that the methods and instruments used for data collection are appropriate and properly calibrated, and that the data collected are fully and thoroughly documented. When developing these procedures, consideration should be given to the number of replicate and repeat measurements needed, as determined by appropriate statistical methods. Consideration should be given to the need for the peer review of data collection activities and results. The data should be traceable to their origin and should be developed into a coherent, well documented description and interpretation of site characteristics. The management system should, as appropriate, include processes and procedures for the qualification of data (see e.g. Ref. [49]) that were not collected in accordance with the management system.

5.162.   The process and procedures for site characterization included in the management system should facilitate the development of the safety case and the conduct of safety assessments, and should allow for the prompt identification of potentially significant gaps in information.

5.163.   The initiation of field based activities at a site — surveys, for example — may heighten the awareness of local people and other interested parties. The process for initiating on-site activities should be carefully planned and implemented. The process should include complying with all applicable regulatory requirements, including giving appropriate notifications to the regulatory body, and engaging with local people and other interested parties, for example on the following:

(a)   Which field based activities could be conducted;
(b)   What the objectives and limits of the field based activities could be;
(c)   How decisions regarding the field based activities could be taken.

5.164.   A carefully designed and agreed site characterization programme (e.g. involving non-intrusive techniques such as surface based geophysics) should be developed and followed in order to develop a sufficient understanding of the baseline conditions (e.g. environmental, hydrogeochemical) at the site before it is subject to significant disturbance by intrusive activities (e.g. the drilling of boreholes, excavation).

Design of radioactive waste management facilities and activities

5.165.   The management system should recognize that the design process for a radioactive waste management facility should be part of a larger iterative process of optimization that involves the development of the safety case for the facility (see para. 5.3 of GSR Part 5 [3] and para. 4.12 of SSR-5 [4]). Site knowledge, facility design and safety arguments and assessments should be refined iteratively to develop a robust safety case and well founded technical specifications to ensure that the facility will be safely constructed, operated and closed or decommissioned, as appropriate. Typically, this proceeds as follows:

(a)   Development of a preliminary, conceptual design for the radioactive waste management facility;
(b)   Assessment of the level of safety that would be provided by the conceptual design for different combinations of waste, facility characteristics, site properties and assumed system performance (e.g. the behaviour of the host rocks for a geological disposal facility);
(c)   Evaluation of the robustness and reliability of the design using the results of the safety assessment and other safety case arguments;
(d)   Refinement, as necessary, and more detailed specification of the design in order to improve safety, environmental protection and the overall feasibility of the design;
(e)   Revision of the safety case using the revised design.

5.166. The optimization process described in para. 5.165 is usually repeated several times until a coherent set of detailed specifications for the facility design and associated safety assessments are obtained and compiled in the safety case. The management system should include processes for appropriate review and approval of the facility design. The operating organization should ensure that there is regular and frequent communication and reporting of progress between the organizations involved in safety assessment and facility design.

5.167. The management system should ensure that the facility includes design features and measures (including inspections, maintenance of structures, systems and components, and monitoring) to optimize protection and safety and waste management activities, and to facilitate operation and closure or decommissioning, as appropriate.

5.168. The management system should include specific procedures for the design of facilities and activities for the management of heat generating waste (including the processing and interim storage of liquid high level waste (see WS-G-6.1 [44], Refs [50–52] and IAEA Safety Standards Series No. SSG-42, Safety of Nuclear Fuel Reprocessing Facilities [53]) and the storage of spent fuel that is considered radioactive waste (see IAEA Safety Standards Series No. SSG-15 (Rev. 1), Storage of Spent Nuclear Fuel [54])). Particular consideration should also be given to the thermal dimensioning of disposal facilities for high level radioactive waste, which involves determining appropriate combinations of waste thermal power, waste package and disposal tunnel spacing, and temperatures, particularly in the engineered barrier system, given the environmental conditions and thermal properties of the disposal site (see Refs [55, 56]).

5.169. The management system should include processes to acquire, review, track, quantify and qualify all design data and to demonstrate their suitability before they are used as input data in any system, computer program or computer model. This includes data from literature searches, laboratory tests, field tests and observations, seismic analyses, monitoring and measurements, and test results from other relevant sources.

5.170. There are always uncertainties associated with data, including data on natural systems and data on engineered structures and components. The management system should ensure that uncertainties in data and the basis for their estimation are clearly documented so that these uncertainties can be taken into account during the facility design and safety assessment process.

5.171. The management system should include a process for ensuring that lessons, knowledge and experience from comparable facilities and projects, including those conducted nationally and internationally, are taken into account at all stages in the design of a radioactive waste management facility.

Construction of radioactive waste management facilities

5.172. The management system should include a process and procedures to ensure that the facility is constructed in accordance with the design, as described in the safety case approved by the regulatory body, the conditions of the authorization and any other relevant requirements (e.g. for environmental protection during construction works) (see Requirement 18 of GSR Part 5 [3] and Requirement 17 of SSR-5 [4]).

5.173. The management system should include the establishment of clear lines of communication between the organizations involved in safety assessment, facility design and construction. Procedures should be put in place for the control and issue of design information and work instructions. The operating organization should ensure that there is regular and frequent communication and reporting of progress between the organizations involved in safety assessment, facility design and construction.

5.174. The management system should include a process and procedures to ensure that, prior to starting construction, the construction organization confirms that the information it has from the design process is up to date and properly informed by the current understanding of site conditions. Procedures should also be included for the gathering of information during construction (e.g. on the nature of the geological formations and their physical–mechanical and hydrogeochemical responses to the construction activities), for the interpretation of this new information and for the updating of the safety case and the facility design, as necessary.

5.175. The management system should include a process and procedures for the operating organization, particularly of disposal facilities, to ensure that the construction works do not unduly disturb the surrounding (e.g. hydrogeochemical) environment. This should include appropriate monitoring to determine the extent of disturbance caused by the construction of the facility. The operator of a geological disposal facility should include in its management system a process and procedures for responding to unexpected rock and groundwater conditions that may be encountered during construction.

5.176.   With regard to the construction of radioactive waste disposal facilities, para. 4.33 of SSR-5 [4] states that "Sufficient flexibility in engineering techniques has to be adopted to allow for variations to be encountered, such as variations in rock conditions or groundwater conditions in underground facilities."

5.177.   The management system should include procedures to demonstrate that any changes to the construction approach, facility design or detailed layout are consistent with safety and are documented together with information on the associated decision making processes.

Operation of radioactive waste management facilities

5.178.   The management system should include a process and procedures to ensure that facilities are operated in accordance with applicable international standards, national regulations, authorization conditions and the design assumptions described in the safety case approved by the regulatory body.

5.179.   The management system should include a process and procedures for the measurement and recording of appropriate data with which to identify and characterize waste at each step in the waste management programme. The process and procedures should ensure that all measurements are made with appropriately calibrated equipment, and that waste items (e.g. individual waste packages) are identified in a unique way and the identification is traceable to the associated records. The procedures should include consideration and specification of the levels of variability and uncertainty in the waste characterization data that are acceptable. Appropriate records should be kept of the inventory of radioactive waste in individual waste packages, particularly in cases where the waste stream might be heterogeneous; these records should focus on waste and radionuclides that are important to safety. The procedures should take account of the need for continued identification and characterization of waste even if a waste item is divided or modified (e.g. repackaged).

5.180.   The management system should include a process and procedures to ensure that it is readily possible to determine the history of a waste item from its documentation. This involves ensuring that information on the nature and history of the waste item is retained and is made available when needed. This is particularly pertinent for radioactive waste that has been emplaced in radioactive waste storage and disposal facilities, especially where the storage conditions are potentially corrosive. The methods used for the physical identification of waste items should be suitably durable. The status of a waste item should be marked either directly on the item, or in documents that are traceable to the item, or both, depending on the circumstances. Consideration should be given to the effects of any marking of waste packages on their degradation. The status of a waste item may in addition be indicated by tags, stamps or other suitable means.

5.181.   The management system should include a process and procedures for the development of waste package specifications for the radiological, physical and chemical characteristics of waste and waste packages. The specifications should also identify which sources may be cleared from regulatory control and which substances may be discharged from waste management facilities.

5.182.   The feasibility of satisfying the waste acceptance criteria in successive waste management steps should be taken into account when defining waste package specifications. The management system should include provisions to ensure that the waste package specifications are consistent with the safety assessments, especially the safety assessments for waste storage and disposal.

5.183.   Specification of waste package characteristics alone might be insufficient, given the impracticality of testing some active waste forms and waste packages. In such cases, the management system should include provisions to ensure that waste specifications also include the composition of feed materials, so that any unexpected variation in the composition of feed materials prompts a reassessment or a non-conformance designation, as appropriate. The management system should also include a procedure for defining the critical operating parameters (e.g. maximum temperatures) of waste processing processes on the basis of the relevant safety assessment.

5.184.   The management system should include a process for ensuring that the relevant organizations are involved in deriving and agreeing specifications for waste and waste packages. These organizations will normally include the following:

(a)   The operating organization of the disposal facility;
(b)   The generator of the waste;
(c)   The owner of the waste (where appropriate);
(d)   The operating organizations of predisposal management facilities;
(e)   The regulatory body.

5.185.   The management system should include a process for ensuring that the specifications for waste and waste packages are used, where relevant, by any organizations that supply services, that manufacture or supply waste containers, or that condition waste.

Waste acceptance at radioactive waste management facilities

5.186.   Waste acceptance criteria that are consistent with the safety case are required to be derived by the operating organization of a radioactive waste management facility (see Requirement 12 of GSR Part 5 [3] and Requirement 20 of SSR-5 [4]). The waste acceptance criteria should also be consistent with other relevant requirements, including those relating to the transport of the waste. The waste acceptance criteria should be discussed with, and explained to, the waste generators and other organizations involved in the management of the waste and should be agreed with the regulatory body.

5.187.   The management system should include procedures for waste acceptance to ensure that the facility only accepts waste that is consistent with the safety case. The management system should include processes and procedures to regularly maintain safety equipment and, in particular, to ensure that equipment used for waste acceptance purposes is suitably calibrated on a regular basis. As stated in para. 4.26 of GSR Part 5 [3], "The operators' procedures for the reception of waste have to contain provisions for safely managing waste that fails to meet the acceptance criteria; for example, by taking remedial actions or by returning the waste."

5.188.   The management system for a storage facility for radioactive waste should include provisions to ensure that prior to storing waste the following are confirmed:

(a)   The waste meets the waste acceptance criteria for the facility.
(b)   The waste is properly identified.
(c)   The required documentation and records are available and acceptable.
(d)   All necessary processes for waste processing have been undertaken and completed satisfactorily.
(e)   The waste and waste packages do not show signs of unacceptable deterioration.
(f)   If applicable, measures for criticality control are in place, are effective and are maintained.
(g)   The intended movements of waste within the storage facility can be performed safely, preclude inadvertent criticality and optimize occupational exposures.
(h)   Procedures are in place for the following:
   (i)   Monitoring the integrity of waste packages;
   (ii)   Monitoring and controlling environmental conditions in the store (e.g. temperature, humidity, ventilation);

(iii) Surveillance of the store and of the status of equipment to facilitate maintenance and replacement, as needed, and surveillance for the detection of unintended events and accidents and mitigation of consequences;

(iv) Ensuring that waste can be readily identified, located and accessed for inspection and retrieval.

(i) Suitable locations and storage capacity exist within the facility.

5.189. The management system for a disposal facility for radioactive waste should include provisions to ensure that prior to emplacing waste the following are confirmed:

(a) The waste meets the waste acceptance criteria for the facility.

(b) The waste is properly identified.

(c) The required documentation and records are available and acceptable.

(d) All necessary processes for waste processing have been undertaken and completed satisfactorily.

(e) The waste and waste packages do not show signs of unacceptable deterioration.

(f) If applicable, measures for criticality control are in place, are effective and are maintained.

(g) Intended movements of waste within the disposal facility can be performed safely, preclude inadvertent criticality and optimize occupational exposures.

(h) Procedures are in place for the following:

(i) Monitoring the integrity of waste and waste packages;

(ii) Monitoring and controlling environmental conditions in the disposal facility (e.g. temperature, humidity, ventilation, rock fall, water inflow);

(iii) Surveillance of the disposal facility and of the status of equipment to facilitate maintenance and replacement, as needed, and surveillance for the detection of unintended events and accidents and mitigation of consequences;

(iv) Ensuring that waste can be readily identified, located and accessed for inspection.

(i) Suitable locations and space exist within the facility for the waste. The management system for geological disposal facilities may also need to include a process and procedures to ensure the suitability of the host rock surrounding the disposal locations (see e.g. Ref. [57]). Such a process might, for example, seek to avoid locations in highly fractured or hydraulically conductive rock.

5.190. The management system should contain procedures to ensure conformance of waste to the relevant waste acceptance criteria for the facility, and this should be independently verified by personnel other than those who prepared the waste and waste packages. The manner in which such verifications are carried out will differ in accordance with the waste and with a graded approach. For example, for low level radioactive waste that can be safely handled manually, verification may consist of directly examining and measuring the characteristics of the individual waste packages. This method is unlikely to be acceptable for intermediate level radioactive waste or high level radioactive waste packages because of the high radiation levels generated. For packages containing waste of these types, verification should be carried out using a combination of indirect methods, such as the following:

(a) Video surveillance of the waste management processes (e.g. waste immobilization by cementation or vitrification, testing of package closure welds);
(b) Sample checks on activities critical to the quality of waste packages (e.g. production of metal used to fabricate metal containers, preparation of concrete for overpacks);
(c) Remote measurement of radiation levels around packages and video surveillance checks of the external condition of packages;
(d) Examination of the data recorded for each waste package.

Waste emplacement and installation of engineered barriers

5.191. The management system for a radioactive waste management facility should include provisions to ensure that waste is emplaced in the facility in accordance with the safety case and in compliance with the authorization for the facility, and that waste emplacement is undertaken in accordance with defined procedures. These procedures should specify how operating personnel should respond to the occurrence of unintended events and accidents, for example equipment failure or the dropping of a waste package.

5.192. The management system for a radioactive waste disposal facility should include processes and procedures to ensure that only appropriate materials are used in constructing engineered barriers and that the engineered barriers are manufactured and emplaced or installed in accordance with the design requirements and the safety case and as approved by the regulatory body. One approach to this is the compilation of relevant information and the definition of 'production lines' for the construction of disposal facility components such as waste packages, buffer, backfill and closure (see e.g. Ref. [58] and references therein), and the implementation of arrangements for the inspection of engineered barriers. These arrangements should address the supply of materials, their quality management, their interim storage under suitable environmental conditions, barrier manufacture and installation, and barrier inspection and testing. The management system should take into account the various restrictions that may be imposed on the manufacture, emplacement and installation of barriers, for example by environmental conditions, interactions between different materials, interactions with other ongoing construction processes, the rate of disposal of radioactive waste and the rate at which engineered barriers need to be installed. The management system should also include procedures to record the quantities of other (i.e. non-radioactive) materials emplaced in a disposal facility so that their possible effects on safety can be assessed.

5.193. Consideration should be given to the demands that will be placed on the structures, systems and components by the conditions that might occur in the facility. Waste stores may experience considerable temperature changes. Radioactive waste disposal facilities might at different times be hot, dry, dusty, humid, wet or cold. Account should also be taken of any restrictions on operations (e.g. due to limited space and accessibility, or high radiation levels).

5.194. The management system should include procedures for fully documenting the inventory of waste received at and emplaced in the facility, including details of the radionuclides and activity levels, relevant properties of the waste forms and the locations of the waste packages emplaced in the facility. The management system should include a process and procedures to ensure that the waste emplacement plans are developed in accordance with the waste acceptance criteria and the assumptions in the safety case (e.g. to prevent undesirable interactions between different wastes, to prevent criticality).

Decommissioning or closure of radioactive waste management facilities

5.195. The management system should include a process and procedures to ensure that radioactive waste management facilities are decommissioned or closed, as appropriate, in accordance with the conditions of the authorization and the relevant decommissioning plan and safety case (see also Requirement 20 of GSR Part 5 [3] (in relation to decommissioning) and Requirement 19 of SSR-5 [4] (in relation to closure)).

5.196. Requirements for the management system for the decommissioning of facilities, including predisposal management facilities, are established in IAEA Safety Standards Series No. GSR Part 6, Decommissioning of Facilities [59]. In particular, Requirement 7 of GSR Part 6 [59] states that "**The licensee shall ensure that its integrated management system covers all aspects of decommissioning.**" Paragraph 4.2 of GSR Part 6 [59] states that "The integrated management system shall enable the planning and implementation of decommissioning actions with the prime goal of ensuring that decommissioning is conducted safely."

5.197. In accordance with paras 4.4 and 4.6 of GSR Part 6 [59], decommissioning is required to be conducted by suitably qualified and experienced personnel and controlled by the use of written procedures.

5.198. The management system should include processes and procedures to ensure that there is traceability for all waste generated, including during decommissioning. This involves maintaining up to date records of the waste generated, stored in the facility and transferred to another authorized facility, specifying the waste quantities, characteristics, treatment methods and destination.

5.199. Requirement 19 of SSR-5 [4] states:

> "**A disposal facility shall be closed in a way that provides for those safety functions that have been shown by the safety case to be important after closure. Plans for closure, including the transition from active management of the facility, shall be well defined and practicable, so that closure can be carried out safely at an appropriate time.**"

5.200. The management system should include plans for the sealing of any preferential pathways (i.e. routes for the migration of radionuclides) that may have been introduced as a result of site characterization or other investigations or of the construction and operation of the disposal facility (e.g. by the drilling of boreholes or the opening of fractures).

5.201. The management system should include processes and procedures to ensure that the disposal system remains safe and that records are adequately maintained after the closure of the facility. Requirement 22 of SSR-5 [4] states:

> "**Plans shall be prepared for the period after closure to address institutional control and the arrangements for maintaining the availability of information on the disposal facility. These plans shall be consistent with passive safety features and shall form part of the safety case on which authorization to close the facility is granted.**"

Monitoring of radioactive waste management facilities

5.202. The management system should include a process and procedures to ensure that facilities are monitored in accordance with the authorization and with the safety case approved by the regulatory body.

5.203. Prior to the construction and operation of a radioactive waste disposal facility (or any associated underground research facility), monitoring should be carried out to gather information and thereby provide a baseline of the existing, undisturbed (e.g. hydrological, geochemical) conditions at the site. Further recommendations on establishing these baseline conditions are provided in IAEA Safety Standards Series No. SSG-31, Monitoring and Surveillance of Radioactive Waste Disposal Facilities [60].

5.204. The management system should also include provisions for the establishment of a monitoring programme to be implemented during the operation of the radioactive waste management facility. This programme should gather information to confirm the safety of workers and members of the public, the protection of the environment and the condition of the facility. This will include, for example, monitoring radiation levels and contamination levels, as well as other parameters, as appropriate, such as ventilation, humidity, groundwater conditions, rock creep and rock stress, and temperature. Monitoring should also be carried out during the operational period to confirm the absence of any conditions that could affect the safety of the site after facility decommissioning or closure, as appropriate. The management system should include procedures, as necessary, for the monitoring of active control systems (e.g. temperature, humidity, ventilation, alarm systems), of waste package integrity, of other relevant equipment (e.g. for the detection and mitigation of accidents) and of the maintenance of waste package identification measures.

5.205. Particular consideration should be given in the management system to the need to develop and implement monitoring programmes for long periods of radioactive waste storage, disposal facility operation and post-closure active institutional control of disposal facilities. These monitoring programmes should be developed by considering the following:

(a) Regulatory requirements;
(b) The objectives of monitoring;
(c) The views of interested parties;
(d) The characteristics of the facility and the site and its surroundings;

(e)	The events and processes that may affect the facility (e.g. earthquakes, corrosion and other waste degradation processes);

(f)	The practicalities of monitoring and the available technologies.

5.206.	The monitoring programmes and their justification should be documented in the safety case.

5.207.	The management system should include procedures for taking actions, as necessary, in response to the results obtained from the monitoring programme and for communicating with interested parties on monitoring results.

5.208.	Further recommendations on monitoring and surveillance of radioactive waste disposal facilities are provided in SSG-31 [60].

6. CULTURE FOR SAFETY

6.1. Requirement 12 of GSR Part 2 [5] states that **"Individuals in the organization, from senior managers downwards, shall foster a strong safety culture. The management system and leadership for safety shall be such as to foster and sustain a strong safety culture."**

6.2. Senior management should be committed to developing a culture for safety, and should communicate this within the organization and demonstrate this through their own actions.

6.3. Paragraph 5.2 of GSR Part 2 [5] states:

"Senior managers and all other managers shall advocate and support the following:

(a)	A common understanding of safety and of safety culture, including: awareness of radiation risks and hazards relating to work and to the working environment; an understanding of the significance of radiation risks and hazards for safety; and a collective commitment to safety by teams and individuals;

(b)	Acceptance by individuals of personal accountability for their attitudes and conduct with regard to safety;

(c)   An organizational culture that supports and encourages trust, collaboration, consultation and communication;

(d)   The reporting of problems relating to technical, human and organizational factors and reporting of any deficiencies in structures, systems and components to avoid degradation of safety, including the timely acknowledgement of, and reporting back of, actions taken;

(e)   Measures to encourage a questioning and learning attitude at all levels in the organization and to discourage complacency with regard to safety;

(f)   The means by which the organization seeks to enhance safety and to foster and sustain a strong safety culture, and using a systemic approach (i.e. an approach relating to the system as a whole in which the interactions between technical, human and organizational factors are duly considered);

(g)   Safety oriented decision making in all activities;

(h)   The exchange of ideas between, and the combination of, safety culture and security culture."

6.4. Managers should also support the identification of relevant actual and potential incidents (including accidents) and non-conformances and be involved in discussions on how these should be rectified and prevented in the future.

6.5. The highest level of documentation in the management system should make leadership for safety the utmost priority, forming a basis for promoting safety culture. The management system documentation should describe the responsibilities of leadership roles (e.g. senior managers, managers) and of the roles of workers for safety and for the development, implementation and fostering of a culture for safety. Internal communication relevant to fostering a culture for safety should cover aspects such as the following:

(a)   Management policy, objectives and strategy;

(b)   The management system documentation;

(c)   Assessments of the culture for safety;

(d)   Processes and procedures for conducting radioactive waste management activities;

(e)   Organizational changes;

(f)   The safety case for the facility and activities, the status of waste management activities, and plans for the future;

(g)   Technical and quality management issues (e.g. problems and their resolution, planned improvements and innovations);

(h) Radiation protection issues (e.g. trends in occupational exposure and in releases to the environment, evaluation of incidents, including accidents);

(i) Regulatory and statutory issues (e.g. the preparation of information to fulfil regulatory requirements and licence conditions, preparation for new laws and requirements on radiation protection and safety, on waste management and on environmental protection).

6.6. A strong culture for safety supports the safe and successful conduct of activities in accordance with the management system. A culture for safety is also an important aspect of organizational effectiveness, safety performance and human performance. A questioning attitude to prevent mistakes, a 'no-blame' attitude (including a commitment to the freedom to express ideas) and self-reflection should be demonstrated by all individuals. The management system should include provisions to ensure that individuals can raise safety issues without fear of penalty, harassment, intimidation, retaliation or discrimination.

6.7. The management system should support the development, implementation and continued enhancement of a strong culture for safety, for example by promoting the adoption of best practices, regardless of the type, scale, complexity, duration and evolution of the waste management activities.

6.8. The management system should contain provisions to support a culture for safety throughout all levels in the organizations involved in the waste management process and for all stages in the lifetime of a waste management facility or activity.

6.9. Senior management should ensure that working conditions and arrangements promote a strong culture for safety and improve the motivation and competence of personnel. The management system should include provisions to ensure that the management and supervision of waste management activities encourage safe ways of working.

6.10. There are specific aspects of radioactive waste management to be taken into account when fostering a culture for safety, as follows:

(a) Individuals should not only consider immediate and short term safety, but should also consider the longer term safety implications of activities, which in some instances might not be manifested until several generations later; the management system should provide individuals with sufficient knowledge to do this. The management system should aim to engender and implement an enduring culture for safety, for example to ensure consistency in the production of high quality waste containers and waste packages, in the monitoring of waste and facility degradation, and in the keeping of records over the potentially very long period of time for which the radioactive waste will remain hazardous.

(b) The waste hierarchy should be applied and the generation of radioactive waste should be minimized.

(c) Where radioactive waste is transferred to other organizations, the safety implications of the actions undertaken at a facility might impact on the receiving organization.

(d) Mistakes in radioactive waste management could lead to non-conforming waste packages, which consequently may have no identified treatment or disposal route; although there might be no immediate safety consequence, a legacy could be left for subsequent generations to manage (see also paras 5.118–5.120).

(e) Personnel, particularly at underground facilities, can sometimes be exposed to non-radiological safety risks that are greater than those posed by radiological hazards. The operating organization should ensure that risks are considered in an integrated manner and that effective overall controls are put in place.

(f) Personnel should use the safety case to determine an appropriate balance between operational safety and post-closure safety at radioactive waste disposal facilities.

7. MEASUREMENT, ASSESSMENT AND IMPROVEMENT OF THE MANAGEMENT SYSTEM

7.1. Requirement 13 of GSR Part 2 [5] states that "**The effectiveness of the management system shall be measured, assessed and improved to enhance safety performance, including minimizing the occurrence of problems relating to safety.**"

7.2. In particular, GSR Part 2 [5] states:

"6.1. The effectiveness of the management system shall be monitored and measured to confirm the ability of the organization to achieve the results intended and to identify opportunities for improvement of the management system.

"6.2. All processes shall be regularly evaluated for their effectiveness and for their ability to ensure safety.

"6.3. The causes of non-conformances of processes and the causes of safety related events that could give rise to radiation risks shall be evaluated and any consequences shall be managed and shall be mitigated. The corrective actions necessary for eliminating the causes of non-conformances, and for preventing the occurrence of, or mitigating the consequences of, similar safety related events, shall be determined, and corrective actions shall be taken in a timely manner. The status and effectiveness of all corrective actions and preventive actions taken shall be monitored and shall be reported to the management at an appropriate level in the organization.

"6.4. Independent assessments and self-assessments of the management system shall be regularly conducted to evaluate its effectiveness and to identify opportunities for its improvement. Lessons and any resulting significant changes shall be analysed for their implications for safety.

"6.5. Responsibility shall be assigned for conducting independent assessments of the management system. The organizations, entities (in-house or external) and individuals assigned such responsibilities shall be given sufficient authority to discharge their responsibilities and shall have direct access to senior management. In addition, individuals conducting independent assessments of the management system shall not be assigned responsibility to assess areas under the responsibility of their own line management.

"6.6. Senior management shall conduct a review of the management system at planned intervals to confirm its suitability and effectiveness, and its ability to enable the objectives of the organization to be accomplished, with account taken of new requirements and changes in the organization.

"6.7. The management system shall include evaluation and timely use of the following:

(a)   Lessons from experience gained and from events that have occurred, both within the organization and outside the organization, and lessons from identifying the causes of events;
(b)   Technical advances and results of research and development;
(c)   Lessons from identifying good practices.

"6.8. Organizations shall make arrangements to learn from successes and from strengths for their organizational development and continuous improvement."

7.3. In cases where radioactive waste has long term safety, societal or economic implications, organizations that were not originally interested parties could, in future, inherit responsibility for managing the waste and the associated facilities. The management system should be sustainable and include provision for its own review in a planned manner to maintain confidence that it will evolve to accommodate changes in management goals, strategies, plans and objectives, in order to meet the needs of future interested parties. Organizations involved in radioactive waste management should establish and implement a formal management review process aimed at the improvement of the management system.

7.4. The evaluation of the effectiveness of processes in accordance with para. 6.2 of GSR Part 2 [5] should address all stages of radioactive waste management, including the transfer of radioactive waste between organizations (see para. 5.45). Planning should be done to ensure that monitoring and measurement of the effectiveness of the management system will be continued, as appropriate, during any long periods of radioactive waste storage, disposal facility operation or institutional control of a disposal facility.

7.5. Self-assessment of the management system should include consideration of the following:

(a) Any changes in organizational structure or in the assignment of responsibilities and financial liabilities that could have an effect on radioactive waste management. This should include a consideration of relevant changes at the national level and, where appropriate, at the international level.
(b) The continuation of assessments over long periods, where necessary, for radioactive waste storage and for the operation of a disposal facility as well as for the post-closure period of institutional control.

7.6. Where assessments and self-assessments are performed in relation to the management system for the predisposal management of radioactive waste, the following aspects should be confirmed:

(a) Process variables and controls have not changed from those established in the original validated processes, as described in the safety case approved by the regulatory body.

(b) Inspections and measurements are being performed in accordance with the management system, and the associated records are being maintained.

(c) The ownership and characteristics of waste are traceable through any transfers of waste, and proper controls are implemented during storage.

(d) The instrumentation used to monitor or control waste management activities has not degraded in service and has not been modified without proper change control.

(e) Critical parameters associated with waste acceptance criteria or other specifications are being controlled within established limits.

(f) Facilities are being operated in accordance with regulatory requirements.

(g) Waste management activities are conducted in a manner that is consistent with the relevant safety assessments.

(h) Waste packages and containers qualified by performance based testing are used within their qualification limits.

(i) Regulatory requirements and conditions in authorizations that relate to waste specifications and waste acceptance criteria have been addressed and are being met.

7.7. Where assessments and self-assessments are performed in relation to the management system for the disposal of radioactive waste, the following aspects should be confirmed:

(a) During the site evaluation stage: Sufficient information is gathered about the nature of the site, including the geological formations, to establish baseline conditions before the site is disturbed by construction activities. All site characterization data can be traced to their origin, and associated uncertainties are adequately described and explained.

(b) During the design stage: The understanding of the site, the design of the facility and the safety case (including the supporting safety assessments) are being developed concurrently in an integrated fashion, and the final descriptions are adequate and mutually consistent.

(c) During the construction stage:
    (i) Sufficient information is being gathered about the disturbance of the site by construction activities, including the response of the geological formation and of the geochemical and geohydrological conditions to any perturbations induced by the construction activities.
    (ii) Construction activities are being carried out in accordance with the safety case and the authorization issued by the regulatory body, and in a manner that facilitates the optimization of the actual facility layout (e.g. in relation to the host geological formation).
    (iii) The construction materials are of the required quality.

      (iv)   Construction works meet the design requirements.

(d)   During the commissioning stage: Activities are being performed in accordance with documentation, appropriate data are being collected and records are being prepared and maintained, and interfaces between commissioning and operations activities are defined.

(e)   During the operational stage:

      (i)   All prerequisites are being met before waste is emplaced (e.g. waste packages are being checked against and are satisfying the acceptance criteria).

      (ii)   Waste is being emplaced in accordance with the safety case and the authorization issued by the regulatory body.

      (iii)   Monitoring is being conducted in accordance with the monitoring programme and the associated records are being maintained, and monitoring instrumentation has not degraded in service and has not been modified without proper change control.

      (iv)   The safety case and safety assessments are being periodically reviewed in a systematic, planned manner and are being updated as necessary in the light of new data, and any necessary actions are being taken to ensure the continued safety of the facility and waste management activities.

      (v)   Any backfilling, sealing and other activities are being carried out in accordance with the safety case and the authorization issued by the regulatory body.

(f)   During the closure and post-closure stages:

      (i)   Backfilling, sealing and other activities and closure of the facility are being carried out in accordance with the safety case and the authorization issued by the regulatory body.

      (ii)   Monitoring is being conducted in accordance with the monitoring programme and the associated records are being maintained, and monitoring instrumentation has not degraded in service and has not been modified without proper change control.

      (iii)   The safety case and safety assessments are being periodically reviewed in a systematic, planned manner and are being updated as necessary in the light of new data, and any necessary actions are being taken to ensure the continued safety of the facility and waste management activities.

      (iv)   Appropriate information on the condition of the radioactive waste disposal facility has been transferred if responsibility for the facility has been transferred.

7.8. Individuals at all levels in the organization should review their work critically to identify areas for improvement. Formal assessments to verify the implementation and effectiveness of the management system may be performed by any of the following:

(a) An organizational unit within the organization itself, which is independent of cost pressure or production pressure, and is independent of the line management responsible for managing and implementing the process being assessed;
(b) Other organizations in the waste management programme;
(c) The regulatory body or other national or local authorities;
(d) International organizations;
(e) A suitably qualified and experienced external organization.

7.9. In conducting reviews of the management system, consideration should be given to whether the structure and content of the management system are still suitable, adequate and effective, especially if the waste management activities continue for a long time or if there is a long period of active institutional control after closure. In such reviews, experience gained from the waste management programme and experience from other facilities and programmes within the State and in other States should be taken into account.

7.10. Reviews of the management system for a radioactive waste management programme should address all aspects of the management system on a periodic basis. The frequency of such reviews should be justified and agreed with the regulatory body, and should take into account the following:

(a) Changes in organizations;
(b) Changes in the governmental, legal and regulatory framework;
(c) Changes in waste management activities;
(d) Significant non-conformances detected;
(e) The need to verify the adequacy of any corrective actions.

7.11. The management system should include a process through which deficiencies are addressed and potential improvements are identified and implemented.

7.12. Reviews of the management system for predisposal management facilities and activities may focus on specific aspects, such as the following:

(a) The specific radioactive waste management activities (e.g. pretreatment, treatment, conditioning, storage) under the control of the organization being assessed;
(b) The safety case and safety assessments for these activities;
(c) The quality of waste packages produced.

7.13. Reviews of the management system for radioactive waste disposal facilities may focus on specific aspects, such as the following:

(a) Site characterization and the disposal concept;
(b) Facility design and safety case development;
(c) Research and development projects and results;
(d) The quality of waste packages and their performance in fulfilling their safety functions;
(e) Specific activities such as excavation, waste emplacement and engineered barrier construction;
(f) Performance of the disposal facility during operation;
(g) Arrangements for closure of the facility and for institutional control;
(h) The safety case;
(i) The performance of the radioactive waste disposal facility as may be determined by monitoring of the disposal system.

7.14. The reviews of the management system should also aim to identify potential non-conformances and recommend actions to prevent their occurrence. This is particularly important when waste management activities are carried out by a number of different organizations, when organizational arrangements change and during long periods of waste storage and disposal facility operation.

7.15. Senior management should support the reviews of management system processes by encouraging the effective identification and correction of non-conformances and monitoring the corrective actions.

7.16. In addition to identifying lessons from operating experience (including near misses) and from incidents (including accidents) (see para. 6.7 of GSR Part 2 [5]), benchmarking by interaction with the operating organizations of other relevant facilities, as appropriate, may also identify potential improvements that warrant consideration.

7.17. Action plans should be developed that identify how, where and when improvements may be made to the management system and to processes. These plans should specify how the improvements will be evaluated to demonstrate that they have been achieved.

7.18. Goals for continuous improvement (e.g. to minimize the number of non-conformances over time) should be embedded in the overall plans and objectives of the organization to demonstrate that continuous improvement is an integral part of normal business and that senior management is fully committed to its success.

7.19. Requirement 14 of GSR Part 2 [5] states that "**Senior management shall regularly commission assessments of leadership for safety and of safety culture in its own organization.**"

7.20. The operating organization should commit to the achievement of high standards of leadership for safety and culture for safety by using self-assessments in which performance is evaluated by reference to internal indicators or by comparison with the performance of other organizations. Self-assessment may involve self-evaluation, self-inspection or self-audit. Senior management is required to ensure that such self-assessment makes use of recognized experts in the assessment of leadership and of safety culture (para. 6.9 of GSR Part 2 [5]). In addition, senior management is required to ensure that an independent assessment of leadership for safety and of safety culture is conducted for enhancement of the organizational culture for safety (para. 6.10 of GSR Part 2 [5]).

7.21. Senior management should make arrangements to measure the effectiveness of leadership and of culture for safety and to demonstrate the performance of managers. Different tools could be used such as surveys, interviews, observations and analysis of the behaviour and achievements of managers.

7.22. Safety performance indicators for leadership for safety and for culture for safety should be developed. The following are examples of such indicators:

(a) The number of safety improvement proposals made, and the percentage of such proposals implemented;
(b) The number of safety inspections conducted by senior management;
(c) The number of safety audit recommendations implemented during a specific period.

7.23. The results of the assessments of leadership for safety and culture for safety, including the degree to which safety performance indicators are met, should be communicated within the organization.

Appendix

ELEMENTS OF THE MANAGEMENT SYSTEM FOR ORGANIZATIONS INVOLVED IN RADIOACTIVE WASTE MANAGEMENT OR ITS REGULATION

A.1. This appendix provides a list of elements of the management system for organizations involved in the management of radioactive waste or its regulation. Not all the elements listed will be relevant to all organizations involved in the management of radioactive waste or its regulatory oversight. In some cases, further processes and procedures may be needed. The precise definitions of, and the boundaries between, the management system elements included in an organization's management system, and the level of detail contained in the processes and procedures, should reflect the nature of the organization concerned and its role and situation, and should be applied according to the graded approach [31].

A.2. Elements relating to the management system include the following:

(a) A process for formally assigning to senior management responsibility for the management system and for its application to achieve the fundamental safety objective.
(b) Descriptions of the structure of the organization and the responsibilities and authorities and decision making powers of individuals for processes.
(c) Processes for ensuring that managers understand the concept of leadership for safety and possess and demonstrate appropriate leadership capabilities.
(d) A process for establishing the management system.
(e) Processes for defining goals, strategies, plans and objectives, consistent with the organization's safety policy.
(f) Processes and procedures for the identification of, interaction with and involvement of interested parties in decision making.
(g) Processes and procedures to ensure that interdependencies between the steps in the radioactive waste management process are considered.
(h) A process for integration of all elements of the management system so that safety is not compromised.
(i) A process for applying and demonstrating the application of the graded approach to the management system commensurate with the risks associated with the facilities and activities being controlled.

(j) Processes and procedures for documentation of the management system. This documentation includes the following:
 (i) Policy statements of the organization on values and behavioural expectations;
 (ii) A statement of the fundamental safety objective;
 (iii) A description of the organization and its structure;
 (iv) A description of responsibilities and accountabilities;
 (v) The levels of authority, including all interactions of those managing, performing and assessing work, for all processes;
 (vi) A description of how the management system complies with regulatory requirements;
 (vii) A description of the interactions with external organizations and with interested parties.
(k) Processes and procedures to ensure that the documents comprising the management system are controlled, usable, readable, clearly identified and readily available at the point of use.
(l) Processes and procedures for the provision of resources (covering both financial and human resources) and their management, including for purchasing and management of the supply chain.
(m) Processes and procedures for recruitment, training and continuing professional development of personnel and for succession planning.
(n) Processes and procedures for the design, management, inspection, testing, verification and validation of processes.
(o) Processes and procedures for the development, documentation, maintenance and use of the safety case for radioactive waste management facilities.
(p) Processes and procedures to ensure the quality of all data, models and safety assessment results.
(q) Processes for developing, promoting the understanding of, communicating and fostering a culture for safety among all personnel of the organization.
(r) Processes and procedures for the steps in the management of radioactive waste (see para. A.4).
(s) Processes and procedures for periodic safety review.
(t) Processes and procedures for measurement, assessment and improvement of the management system.

A.3. Topics that should be considered when developing the management system include the following:

(a) The development and publication of documents;
(b) Change control;
(c) Communication;

(d) Transparency;

(e) Traceability;

(f) Research and development;

(g) Treatment of uncertainties;

(h) Processes and procedures for establishing an integrated national radioactive waste management strategy and programme;

(i) Optimization of the radioactive waste management system, of radioactive waste management facilities and activities, and of protection and safety;

(j) Knowledge management;

(k) Record keeping and archiving of relevant information.

A.4. Elements relating to the steps in radioactive waste management include the following, as appropriate:

(a) Processes and procedures for establishing and maintaining the inventory of radioactive waste;

(b) Processes and procedures for waste characterization and for the recording of relevant information;

(c) Processes and procedures for the clearance of materials and classification of waste;

(d) Processes and procedures for identifying appropriate predisposal management steps and disposal routes for radioactive waste;

(e) Processes and procedures for the siting of facilities;

(f) Processes and procedures for site selection and site characterization;

(g) Procedures for the design of radioactive waste management processes;

(h) Processes and procedures for the design of waste management facilities;

(i) Processes and procedures for controlling the construction of radioactive waste management facilities;

(j) Processes and procedures for the maintenance of facilities and equipment;

(k) Processes and procedures for establishing arrangements for preparedness and response for a nuclear or radiological emergency;

(l) Processes and procedures for controlling the commissioning of radioactive waste management facilities and activities;

(m) Processes and procedures for controlling the operation of radioactive waste management facilities and activities, addressing issues such as facility housekeeping and cleanliness;

(n) Processes and procedures for dealing with unexpected occurrences such as incidents, including accidents;

(o) Processes and procedures for establishing waste package specifications for radioactive waste (e.g. covering waste containers and outputs from waste conditioning);

(p)   Processes and procedures for establishing waste acceptance criteria for radioactive waste management facilities;

(q)   Processes and procedures for waste acceptance at facilities, including compliance checking with the waste acceptance criteria;

(r)   Processes and procedures for dealing with non-conformances, such as cases where waste acceptance criteria are not met;

(s)   Processes and procedures for waste storage, for the control of environmental conditions in storage facilities, for the identification and inspection of stored waste and for the retrieval of waste from storage;

(t)   Processes and procedures for waste emplacement and the installation of engineered barriers in disposal facilities in accordance with the safety case and authorization for the facility;

(u)   Processes and procedures for controlling the decommissioning and closure, as appropriate, of radioactive waste management facilities;

(v)   Processes and procedures for monitoring of radioactive waste management facilities and activities and for monitoring of the environment;[15]

(w)   Processes and procedures for identifying and prioritizing research and development needs and activities to fill gaps in knowledge that are important to safety;

(x)   Processes and procedures for ensuring compliance with regulatory requirements (see e.g. Ref. [61]).

A.5. Elements relating to regulatory functions and processes for radioactive waste management facilities and activities (see also IAEA Safety Standards Series No. GSG-13, Functions and Processes of the Regulatory Body for Safety [62]) include the following:

(a)   Processes and procedures for the authorization of radioactive waste management facilities and activities;

(b)   Processes and procedures for the inspection of radioactive waste management facilities and activities;

(c)   Processes and procedures for the development of regulations and regulatory guidance on radioactive waste management;

(d)   Processes and procedures for review and assessment (e.g. of authorization applications, safety cases and safety assessments for radioactive waste management facilities and activities).

_____

[15] For example, monitoring of environmental conditions in the facility, monitoring of the condition and integrity of waste packages, monitoring of equipment, monitoring of occupational exposures, managing and responding to monitoring data, communicating with interested parties on the monitoring programme and the results obtained.

REFERENCES

[1]   INTERNATIONAL ATOMIC ENERGY AGENCY, IAEA Safety Glossary: Terminology Used in Nuclear Safety and Radiation Protection, 2018 Edition, IAEA, Vienna (2019).

[2]   EUROPEAN ATOMIC ENERGY COMMUNITY, FOOD AND AGRICULTURE ORGANIZATION OF THE UNITED NATIONS, INTERNATIONAL ATOMIC ENERGY AGENCY, INTERNATIONAL LABOUR ORGANIZATION, INTERNATIONAL MARITIME ORGANIZATION, OECD NUCLEAR ENERGY AGENCY, PAN AMERICAN HEALTH ORGANIZATION, UNITED NATIONS ENVIRONMENT PROGRAMME, WORLD HEALTH ORGANIZATION, Fundamental Safety Principles, IAEA Safety Standards Series No. SF-1, IAEA, Vienna (2006).

[3]   INTERNATIONAL ATOMIC ENERGY AGENCY, Predisposal Management of Radioactive Waste, IAEA Safety Standards Series No. GSR Part 5, IAEA, Vienna (2009).

[4]   INTERNATIONAL ATOMIC ENERGY AGENCY, Disposal of Radioactive Waste, IAEA Safety Standards Series No. SSR-5, IAEA, Vienna (2011).

[5]   INTERNATIONAL ATOMIC ENERGY AGENCY, Leadership and Management for Safety, IAEA Safety Standards Series No. GSR Part 2, IAEA, Vienna (2016).

[6]   Joint Convention on the Safety of Spent Fuel Management and on the Safety of Radioactive Waste Management, INFCIRC/546, IAEA, Vienna (1997).

[7]   INTERNATIONAL ATOMIC ENERGY AGENCY, Guidance on the Management of Disused Radioactive Sources, 2018 Edition, IAEA, Vienna (2018).

[8]   INTERNATIONAL ATOMIC ENERGY AGENCY, Regulations for the Safe Transport of Radioactive Material, 2018 Edition, IAEA Safety Standards Series No. SSR-6 (Rev. 1), IAEA, Vienna (2018).

[9]   INTERNATIONAL ATOMIC ENERGY AGENCY, The Management System for the Safe Transport of Radioactive Material, IAEA Safety Standards Series No. TS-G-1.4, IAEA, Vienna (2008).

[10]   INTERNATIONAL ATOMIC ENERGY AGENCY, Classification of Radioactive Waste, IAEA Safety Standards Series No. GSG-1, IAEA, Vienna (2009).

[11]   INTERNATIONAL ATOMIC ENERGY AGENCY, Decommissioning of Nuclear Power Plants, Research Reactors and Other Nuclear Fuel Cycle Facilities, IAEA Safety Standards Series No. SSG-47, IAEA, Vienna (2018).

[12]   INTERNATIONAL ATOMIC ENERGY AGENCY, Decommissioning of Medical, Industrial and Research Facilities, IAEA Safety Standards Series No. SSG-49, IAEA, Vienna (2019).

[13]   INTERNATIONAL ATOMIC ENERGY AGENCY, UNITED NATIONS ENVIRONMENT PROGRAMME, Regulatory Control of Radioactive Discharges to the Environment, IAEA Safety Standards Series No. GSG-9, IAEA, Vienna (2018).

[14]   OECD NUCLEAR ENERGY AGENCY, Geological Disposal of Radioactive Waste: National Commitment, Local and Regional Involvement, NEA Report No. 7082, OECD, Paris (2012).

[15] TAKEUCHI, M.R.H., HASEGAWA, T., HARDIE, S.M.L., McKINLEY, L.E., ISHIHARA, K.N., Leadership for management of high-level radioactive waste in Japan, Environ. Geotech. 7 2 (2020).

[16] INTERNATIONAL ATOMIC ENERGY AGENCY, The Safety Case and Safety Assessment for the Predisposal Management of Radioactive Waste, IAEA Safety Standards Series No. GSG-3, IAEA, Vienna (2013).

[17] INTERNATIONAL ATOMIC ENERGY AGENCY, The Safety Case and Safety Assessment for the Disposal of Radioactive Waste, IAEA Safety Standards Series No. SSG-23, IAEA, Vienna (2012).

[18] INTERNATIONAL ATOMIC ENERGY AGENCY, Governmental, Legal and Regulatory Framework for Safety, IAEA Safety Standards Series No. GSR Part 1 (Rev. 1), IAEA, Vienna (2016).

[19] FOOD AND AGRICULTURE ORGANIZATION OF THE UNITED NATIONS, INTERNATIONAL ATOMIC ENERGY AGENCY, INTERNATIONAL CIVIL AVIATION ORGANIZATION, INTERNATIONAL LABOUR ORGANIZATION, INTERNATIONAL MARITIME ORGANIZATION, INTERPOL, OECD NUCLEAR ENERGY AGENCY, PAN AMERICAN HEALTH ORGANIZATION, PREPARATORY COMMISSION FOR THE COMPREHENSIVE NUCLEAR-TEST-BAN TREATY ORGANIZATION, UNITED NATIONS ENVIRONMENT PROGRAMME, UNITED NATIONS OFFICE FOR THE COORDINATION OF HUMANITARIAN AFFAIRS, WORLD HEALTH ORGANIZATION, WORLD METEOROLOGICAL ORGANIZATION, Preparedness and Response for a Nuclear or Radiological Emergency, IAEA Safety Standards Series No. GSR Part 7, IAEA, Vienna (2015).

[20] FOOD AND AGRICULTURE ORGANIZATION OF THE UNITED NATIONS, INTERNATIONAL ATOMIC ENERGY AGENCY, INTERNATIONAL CIVIL AVIATION ORGANIZATION, INTERNATIONAL LABOUR OFFICE, INTERNATIONAL MARITIME ORGANIZATION, INTERPOL, OECD NUCLEAR ENERGY AGENCY, UNITED NATIONS OFFICE FOR THE COORDINATION OF HUMANITARIAN AFFAIRS, WORLD HEALTH ORGANIZATION, WORLD METEOROLOGICAL ORGANIZATION, Arrangements for the Termination of a Nuclear or Radiological Emergency, IAEA Safety Standards Series No. GSG-11, IAEA, Vienna (2018).

[21] INTERNATIONAL ATOMIC ENERGY AGENCY, Management of Large Volumes of Waste Arising in a Nuclear or Radiological Emergency, IAEA-TECDOC-1826, IAEA, Vienna (2017).

[22] INTERNATIONAL ORGANIZATION FOR STANDARDIZATION, Quality Management Systems — Requirements, ISO 9001:2015, ISO, Geneva (2015).

[23] INTERNATIONAL ORGANIZATION FOR STANDARDIZATION, Environmental Management Systems — Requirements with Guidance for Use, ISO 14001:2015, ISO, Geneva (2015).

[24] INTERNATIONAL ORGANIZATION FOR STANDARDIZATION, Occupational Health and Safety Management Systems — Requirements with Guidance for Use, ISO 45001:2018, ISO, Geneva (2018).

[25] LLW REPOSITORY LIMITED NATIONAL WASTE PROGRAMME, RADIOACTIVE WASTE MANAGEMENT LIMITED STRATEGIC WASTE PROGRAMME, UK Management of Radioactive Waste: An Introductory Good Practice Guide for the Application of the Waste Hierarchy, NDA Report No. NWP/REP/077, Issue 3, Nuclear Decommissioning Authority, UK (2021).

[26] RADIOACTIVE WASTE MANAGEMENT LIMITED, Societal Aspects of Geological Disposal, RWMD Report No. RWM007420, Issue 5, Nuclear Decommissioning Authority, UK (2016).

[27] INTERNATIONAL ATOMIC ENERGY AGENCY, Communication and Consultation with Interested Parties by the Regulatory Body, IAEA Safety Standards Series No. GSG-6, IAEA, Vienna (2017).

[28] KINGDOM OF BELGIUM, National Programme for the Management of Spent Fuel and Radioactive Waste, Directorate General for Energy, Brussels (2015) [English courtesy translation: ONDRAF/NIRAS, Brussels (2015)].

[29] LANDAIS, P., PETIT, L., "Optimization of post dismantling radwaste management: A French innovative initiative – 18013", Proc. WM2018: Annual Waste Management Conf., Phoenix, 2018, Waste Management Symposia, Tempe (2018).

[30] NUCLEAR DECOMMISSIONING AUTHORITY, Geological Disposal — Upstream Optioneering: Summary of LLW/ILW Opportunities, 2015, NDA Report No. NDA/RWM/134, NDA, Harwell, UK (2015).

[31] INTERNATIONAL ATOMIC ENERGY AGENCY, Use of a Graded Approach in the Application of the Management System Requirements for Facilities and Activities, IAEA-TECDOC-1740, IAEA, Vienna (2014).

[32] INTERNATIONAL ATOMIC ENERGY AGENCY, Application of the Management System for Facilities and Activities, IAEA Safety Standards Series No. GS-G-3.1, IAEA, Vienna (2006).

[33] OECD NUCLEAR ENERGY AGENCY, Foundations and Guiding Principles for the Preservation of Records, Knowledge and Memory Across Generations: A Focus on the Post-Closure Phase of Geological Repositories, OECD, Paris (2014).

[34] BUDZIER, A., FLYVBJERG, B., GARAVAGLIA, A., LEED, A., Quantitative Cost and Schedule Risk Analysis of Nuclear Waste Storage (December 10, 2018). Available at SSRN: https://dx.doi.org/10.2139/ssrn.3303410.

[35] EUROPEAN COMMISSION, FOOD AND AGRICULTURE ORGANIZATION OF THE UNITED NATIONS, INTERNATIONAL ATOMIC ENERGY AGENCY, INTERNATIONAL LABOUR ORGANIZATION, OECD NUCLEAR ENERGY AGENCY, PAN AMERICAN HEALTH ORGANIZATION, UNITED NATIONS ENVIRONMENT PROGRAMME, WORLD HEALTH ORGANIZATION, Radiation Protection and Safety of Radiation Sources: International Basic Safety Standards, IAEA Safety Standards Series No. GSR Part 3, IAEA, Vienna (2014).

[36] INTERNATIONAL ATOMIC ENERGY AGENCY, Predisposal Management of Radioactive Waste from Nuclear Power Plants and Research Reactors, IAEA Safety Standards Series No. SSG-40, IAEA, Vienna (2016).

[37] INTERNATIONAL ATOMIC ENERGY AGENCY, Predisposal Management of Radioactive Waste from Nuclear Fuel Cycle Facilities, IAEA Safety Standards Series No. SSG-41, IAEA, Vienna (2016).

[38] INTERNATIONAL ATOMIC ENERGY AGENCY, The Management System for Nuclear Installations, IAEA Safety Standards Series No. GS-G-3.5, IAEA, Vienna (2009).

[39] INTERNATIONAL ATOMIC ENERGY AGENCY, Organization, Management and Staffing of the Regulatory Body for Safety, IAEA Safety Standards Series No. GSG-12, IAEA, Vienna (2018).

[40] SVENSK KÄRNBRÄNSLEHANTERING, RD&D Programme 2019: Programme for Research, Development and Demonstration of Methods for the Management and Disposal of Nuclear Waste, SKB Technical Report No. TR-19-24, SKB, Stockholm (2019).

[41] OECD NUCLEAR ENERGY AGENCY, Management of Uncertainty in Safety Cases and the Role of Risk (Proc. Workshop, Stockholm, 2004), OECD, Paris (2004).

[42] NUCLEAR DECOMMISSIONING AUTHORITY, Geological Disposal: Methods for Management and Quantification of Uncertainty, NDA Report No. NDA/RWM/153, Harwell, UK (2017).

[43] INTERNATIONAL ATOMIC ENERGY AGENCY, Predisposal Management of Radioactive Waste from the Use of Radioactive Material in Medicine, Industry, Agriculture, Research and Education, IAEA Safety Standards Series No. SSG-45, IAEA, Vienna (2019).

[44] INTERNATIONAL ATOMIC ENERGY AGENCY, Storage of Radioactive Waste, IAEA Safety Standards Series No. WS-G-6.1, IAEA, Vienna (2006).

[45] INTERNATIONAL ATOMIC ENERGY AGENCY, Borehole Disposal Facilities for Radioactive Waste, IAEA Safety Standards Series No. SSG-1, IAEA, Vienna (2009). (A revision of this publication is in preparation.)

[46] INTERNATIONAL ATOMIC ENERGY AGENCY, Geological Disposal Facilities for Radioactive Waste, IAEA Safety Standards Series No. SSG-14, IAEA, Vienna (2011).

[47] INTERNATIONAL ATOMIC ENERGY AGENCY, Near Surface Disposal Facilities for Radioactive Waste, IAEA Safety Standards Series No. SSG-29, IAEA, Vienna (2014).

[48] INTERNATIONAL ATOMIC ENERGY AGENCY, UNITED NATIONS ENVIRONMENT PROGRAMME, Prospective Radiological Environmental Impact Assessment for Facilities and Activities, IAEA Safety Standards Series No. GSG-10, IAEA, Vienna (2018).

[49] YAMADA, S., KUNIMARU, T., OTA, T., VOMVORIS, S., GIROUD, N., "Data qualification methodology in the literature survey stage", Proc. 6th East Asia Forum on Radwaste Management Conf., Osaka, 2017, Atomic Energy Society of Japan, Tokyo (2017).

[50] DOBSON, A.J., PHILLIPS, C., "High level waste processing in the U.K. — Hard won experience that can benefit U.S. nuclear cleanup work", Proc. Waste Management (WM'06) Symp., Tucson, 2006.

[51] OFFICE FOR NUCLEAR REGULATION, The Storage of Liquid High Level Waste at Sellafield: Revised Regulatory Strategy (2011),
http://www.onr.org.uk/halstock-sellafield-public.pdf

[52] NATIONAL RESEARCH COUNCIL, Tank Waste Retrieval, Processing, and On-site Disposal at Three Department of Energy Sites: Final Report, The National Academies Press, Washington, DC (2006).

[53] INTERNATIONAL ATOMIC ENERGY AGENCY, Safety of Nuclear Fuel Reprocessing Facilities, IAEA Safety Standards Series No. SSG-42, IAEA, Vienna (2017).

[54] INTERNATIONAL ATOMIC ENERGY AGENCY, Storage of Spent Nuclear Fuel, IAEA Safety Standards Series No. SSG-15 (Rev. 1), IAEA, Vienna (2020).

[55] MYERS, S., HOLTON, D., HOCH, A., Thermal dimensioning to determine acceptable waste package loading and spatial configurations of heat-generating waste packages, Mineral. Mag. **79** 6 (2015) 1625–1632.

[56] IKONEN, K., KUUTTI, J., RAIKO, H., Thermal Dimensioning of Olkiluoto Repository — 2018 Update, Working Report 2018-26, Posiva Oy, Eurajoki, Finland (2018).

[57] McEWEN, T., ARO, S., MATTILA, J., PERE, T., KÄPYAHO, A., HELLÄ, P., Rock Suitability Classification — RSC 2012, POSIVA 2012-24, Posiva Oy, Eurajoki, Finland (2012).

[58] SVENSK KÄRNBRÄNSLEHANTERING, Design and Production of the KBS-3 Repository, SKB Technical Report No. TR-10-12, SKB, Stockholm (2010).

[59] INTERNATIONAL ATOMIC ENERGY AGENCY, Decommissioning of Facilities, IAEA Safety Standards Series No. GSR Part 6, IAEA, Vienna (2014).

[60] INTERNATIONAL ATOMIC ENERGY AGENCY, Monitoring and Surveillance of Radioactive Waste Disposal Facilities, IAEA Safety Standards Series No. SSG-31, IAEA, Vienna (2014).

[61] OECD NUCLEAR ENERGY AGENCY, Managing Information and Requirements in Geological Disposal Programmes, NEA Report No. NEA/RWM/R(2018)2, OECD, Paris (2018).

[62] INTERNATIONAL ATOMIC ENERGY AGENCY, Functions and Processes of the Regulatory Body for Safety, IAEA Safety Standards Series No. GSG-13, IAEA, Vienna (2018).

CONTRIBUTORS TO DRAFTING AND REVIEW

Arvidsson, P.	Swedish Nuclear Fuel and Waste Management Company, Sweden
Bennett, D.G.	International Atomic Energy Agency
Boydon, F.	Consultant, United Kingdom
Carlton, P.	Radioactive Waste Management, United Kingdom
Codée, H.	Central Organization for Radioactive Waste, Netherlands
Delaney, B.	Consultant, United Kingdom
Faß, T.	Gesellschaft für Anlagen- und Reaktorsicherheit (GRS), Germany
Fokke, N.	Central Organization for Radioactive Waste, Netherlands
Hedberg, B.	Swedish Radiation Safety Authority, Sweden
Kumano, Y.	International Atomic Energy Agency
Mingrone, G.	Società Gestione Impianti Nucleari (Sogin), Italy
Mononen, J.	Radiation and Nuclear Safety Authority, Finland
Papaz, D.	Canadian Nuclear Safety Commission, Canada
Yoshida, M.	Nuclear Safety Technology Center, Japan

IAEA
International Atomic Energy Agency

ORDERING LOCALLY

IAEA priced publications may be purchased from the sources listed below or from major local booksellers.

Orders for unpriced publications should be made directly to the IAEA. The contact details are given at the end of this list.

NORTH AMERICA

Bernan / Rowman & Littlefield
15250 NBN Way, Blue Ridge Summit, PA 17214, USA
Telephone: +1 800 462 6420 • Fax: +1 800 338 4550
Email: orders@rowman.com • Web site: www.rowman.com/bernan

REST OF WORLD

Please contact your preferred local supplier, or our lead distributor:

Eurospan Group
Gray's Inn House
127 Clerkenwell Road
London EC1R 5DB
United Kingdom

Trade orders and enquiries:
Telephone: +44 (0)176 760 4972 • Fax: +44 (0)176 760 1640
Email: eurospan@turpin-distribution.com

Individual orders:
www.eurospanbookstore.com/iaea

For further information:
Telephone: +44 (0)207 240 0856 • Fax: +44 (0)207 379 0609
Email: info@eurospangroup.com • Web site: www.eurospangroup.com

Orders for both priced and unpriced publications may be addressed directly to:
Marketing and Sales Unit
International Atomic Energy Agency
Vienna International Centre, PO Box 100, 1400 Vienna, Austria
Telephone: +43 1 2600 22529 or 22530 • Fax: +43 1 26007 22529
Email: sales.publications@iaea.org • Web site: www.iaea.org/publications